BT COMMUNICATIONS TECHNOLOGY SERIES 7

Telecommunications Performance Engineering

Other volumes in this series:

Telecommunications Performance Engineering

Edited by
Roger Ackerley

The Institution of Electrical Engineers

Published by: The Institution of Electrical Engineers, London,
United Kingdom

© 2003: British Telecommunications plc

British Library Cataloguing in Publication Data

A catalogue record for this product is available from the British Library

ISBN 0 86341 341 2

Typeset in the UK by Bowne Global Solutions Ltd, Ipswich
Printed in the UK by T J International, Padstow, Cornwall

CONTENTS

PREFACE

With the consolidation of the communications technology industry, there is renewed focus on cost reduction, maximising usage of resources and delivering improved quality-of-service to the customer. At the same time great technological change is producing an explosion of new applications, and therefore new performance problems to be solved. Furthermore, competitive and regulatory pressures on the industry continue to rise. Risk management is now key as technical and commercial risks are balanced against the need to meet market opportunities on time and to budget. These technology, market, business and customer service drivers are all increasing the need for performance engineering.

Whenever resources are shared there are performance questions to be answered and a performance engineering job to be done. Typical questions are:

- How can the solution be designed to meet these delay and loss requirements?

- How does it scale and what are the 'break points'?

- Where are the potential bottle-necks?

- What control system is required to optimise performance when the platform is overloaded?

- How do we ensure that the solution performs as designed when it goes live?

- How will performance and capacity be managed and what measurements are required to do this?

The aim of this book is to provide an insight into the rich diversity of techniques, tools and knowledge used to address these types of question. The chapters cover the whole of the life cycle from design through to operation of both networks and systems and are written by technical experts currently actively working in the respective areas.

The subject of performance engineering is divided into five parts:

- introduction;

- modelling, measurement and testing;

- overload control;

- in-life performance;

- epilogue.

The first part provides an introduction to the subject of performance engineering by means of an historical perspective covering more than two decades given in Chapter 1. Over this time applications have changed out of all recognition — for example, from designing gradings for electromechanical exchanges to ensuring quality-of-service in IP networks. Perhaps surprisingly, some techniques are as relevant today as they were in the 1970s, while a productive relationship between academics and practitioners continues to produce new methods. This chapter illustrates what drives performance engineering work and explains why it still exists as a discipline today and is needed even more than in the past.

The second part describes the essential performance engineering techniques of modelling, measurement and testing. Performance modelling provides the ability to predict performance through a quantitative understanding of how varying demand affects the behaviour of a system or network. Without it, there is no understanding and merely the process-driven reactions to faults and problems resulting in endless, haphazard experiments to provide a fix. An overview together with the related subject of traffic characterisation is presented in Chapter 2.

The interaction between new services and technologies and existing platforms will provide many performance engineering challenges over the next few years. The rapid introduction of widespread Internet access in the UK through IP-dial required some re-engineering of the PSTN based on the modelling described in Chapter 3. There are critical IP network performance issues remaining to be resolved, and this area therefore continues to be an active area for the application of both analytical and simulation modelling. Chapter 4 provides examples of how modelling is used to provide an understanding of quality of service in IP networks, and how measurement is used as a complementary technique to validate modelling and discover new factors that have not previously been considered.

As shown in Chapter 5, modelling can also be used as a decision support tool for in-life systems. A case-study is presented in Chapter 6 illustrating the role of testing in ensuring that systems meet their performance requirements when they go live. Performance testing is becoming more critical as solutions are developed through the integration of separate application components, and the current economic pressures tend to push the identification and resolution of performance issues further down the supplier chain. This could be despite performance modelling earlier in the life cycle, or perhaps because there wasn't any! Testing is usually the last opportunity to address these before they become operational issues with consequentially higher failure costs.

The third part concerns the design of overload control. There are times such as during civil emergencies, media-stimulated events, adverse natural conditions or failures when demand for resources is considerably greater than normal. The simplest solution is to over-provide capacity to meet the greatest possible demand, but this is usually uneconomic. An alternative is to provide effective overload control, which is perhaps one of the most complex areas of design and therefore often with long development times. Chapter 7 introduces a generic network solution, while Chapter 8 gives an example of a successful implementation.

The fourth part relates to the in-life performance of networks and systems: Starting with Capacity Planning, Chapter 9 shows how performance engineering techniques can be employed to address the challenge of planning large IP networks to provide the required performance and Chapter 10 explains the methodology developed over several years to manage the capacity of a sophisticated platform employed to deliver telemarketing services. Then the concept of a 'Performance Health Check' is portrayed in Chapter 11. The described high-level techniques are used primarily to identify performance risks and issues.

The final part, Chapter 12, provides a futurologist's view on the performance implications of delivering broadband services — a fitting epilogue.

I would like to thank my colleagues who have contributed enthusiastically as authors, reviewers and editors to produce this book. I hope you enjoy reading it and gain a greater understanding of performance engineering and its relevance to the world of communications and beyond.

Roger Ackerley
Critical Solutions Performance,
BT Exact
roger.ackerley@bt.com

CONTRIBUTORS

C G Baisden, Intelligence and Voice Integration, BT Exact, Adastral Park

S F Carter, IP Network Development, BT Exact, Adastral Park

D J Chown, Systems Performance, BT Exact, Adastral Park

J Graham, Network Control Performance, BT Exact, Adastral Park

L G Kirby, Systems Performance, BT Exact, Adastral Park

N W Macfadyen, Network Performance Engineering, BT Exact, Adastral Park

G A R Martin, Design and Performance, BT Exact, Adastral Park

I D Pearson, Futurologist, BT Exact, Adastral Park

W M Pryke, Service Innovation, BT Exact, Adastral Park

J C C Shaw, Applications Performance, BT Exact, Adastral Park

P Singleton, Systems Performance BT Exact, Adastral Park

R L Southgate, IP Network Development, BT Exact, Adastral Park

N W Stewart, Network Performance, BT Exact, Adastral Park

C A van Eijl, Internetworking Performance, BT Exact, Adastral Park

M J Whitehead, Performance Engineering, BT Exact, Adastral Park

P M Williams, Network Control Performance, BT Exact, Adastral Park

1

PERFORMANCE — A RETROSPECTIVE VIEW

N W Macfadyen

1.1   Introduction

Forty years ago — or even thirty — performance engineering was a very different area from what it is now. For a start, it was then termed 'teletraffic'; more significantly, the range of systems and problems studied was much narrower, the variety much less, and the speed of evolution incomparably lower.

It was also focused, to a degree which is difficult to understand from today's perspective. Problems were thoroughly — indeed exhaustively — studied, irrespective of the wider context, until the definitive answer was obtained. Massive simulations of entire switching systems were run and re-run until it was identified whether their capacity at some essentially arbitrary performance level was 2000E or 2001E; and the community of telecommunications engineers, if not that of teletraffic engineers, believed that it mattered.

Only with the opening-up of the business following privatisation and liberalisation in the 1980s did this attitude begin to alter. Within the UK, a sea-change was effected by the so-called 'overall grade-of-service' studies carried out within BT, which made evident for the first time the extent of variability of performance figures over a properly managed network (including how very far the typical system was from its notional design-date target!). Differences which were previously seen as significant, albeit not important, were suddenly placed in context; and the definitive study was never the same again.

That performance engineers, unlike ammonites, still survive in what is a very different world is due to two factors — the importance of the subject, and the skill of its practitioners in handling evolution. It has also been undeniably helped by the symbiotic relationship between practising engineers and academics (generally mathematicians). Telecommunications problems give rise to real examples where mathematical and statistical analysis can be applied, and results obtained and even validated against voluminous measurements — a situation sufficiently rare that it provides academics with both challenges and opportunities. The cross-fertilisation between theory and engineering is of considerable mutual benefit, even though the ultimate aims of the two communities are very different.

1.2   Performance as It Was

When the BT's Teletraffic Division was first formed, networks were primitive. They consisted to all intents and purposes of collections of single circuits, with a combination of Strowger and crossbar switching. Heavyweight processes were in place which gave — or purported to give — definitive forecasts of traffic levels at fixed design dates; and the function of teletraffic was to ensure that the circuit quantities and switches were engineered adequately.

Control was electromechanical, and therefore with limitations; and much work was devoted worldwide to the study of gradings, or limited-accessibility crossconnection patterns. For practical reasons, not every incoming circuit in a large group could see every outgoing circuit, and elaborate schemes were therefore developed to provide fair and optimal patterns to minimise the limitations. BT, in common with many other PTTs, standardised on O'Dell gradings; and teletraffic did much work on improvements to these, such as partially skipped gradings, before two things became apparent:

- firstly, that implementation of any change would cost more than it was worth;
- secondly, that, with the advent of processor-controlled systems, the whole grading technology had become obsolete anyway.

Teletraffic moved on. Strowger switching gave way to crossbar, and worldwide attention settled on link systems — preferably multistage, with up to perhaps seven stages. A variety of more-or-less elegant techniques and approximations arose to analyse these; and the theory reached its final flowering in the early 1970s, with the development of Jacobaeus and other methods, the study of nonblocking and rearrangeable networks, and the elegant but difficult mathematical Takagi graphs. Little of this edifice now remains — Clos networks alone are still found among the desert sands.

The evolution of big complex switches now resulted in the arrival of big complex simulation programs to study their capacity and control; and it was therefore now that the first mainframe time-sharing computers were used in earnest in performance studies. Useful simulation languages were almost non-existent, let alone packages; and, for many years, elaborate simulations were coded in general-purpose languages such as PL/1, with the aid of a few specially written assembly-language routines to handle random-number generation, event-list processing, and similar dedicated areas. Despite (or perhaps because of) this, they were fast, flexible, and productive, even though the machine speeds of the time meant that the actual runs took several hours and were performed overnight.

An area which was not much in evidence at this era was formal quality. A variety of informal checks would be carried out on programs and their output; but testing was decidedly exiguous, and teletraffic engineers piqued themselves that it was their own special skill that allowed them to judge the correctness of the results. In any case, there was always substantial random variation, and any anomalies could be plausibly ascribed to that. A number of weighty reports had to be updated 'with even more accurate modelling' when curious effects like negative intermediate probabilities came to light within the simulations. It is not clear that these (generally small) corrections were ever noticed, let alone applied, by anyone other than their begetters; but the lesson, that thorough review and testing is as much of a *sine qua non* for theoretical studies as for actual engineered plant, was learned by the teletraffic fraternity perhaps rather earlier than in much of the rest of the telecommunications industry.

1.3   Towards Modernity

The next major innovation was that of switches which were actually processor systems themselves. An entire, new area of problems now arose — those of software — and it was some years before it was widely appreciated outside the teletraffic community that these problems existed and were actually important: statements from high levels of management that 'software doesn't have performance problems' required real effort to overcome. Studies of major switches were now focused upon processing load and memory requirements, for memory was very expensive and was strictly limited; and they had to incorporate models of the switch's own self-monitoring and overload control mechanisms. The effort taken to reject a call during congestion became significant; and for the first time the real possibility arose of an exchange crash caused by load, not by hardware failure.

The fresh environment resulted in a prolific development of queueing-theory models applied to telecommunications systems. A wide variety of service-time requirements, arrival processes, and service disciplines were studied; and no conference was complete without a session devoted to theoretical papers on specific queueing systems. It is sad to report that few of these have survived, or proved to have very much relevance or application today.

Data switching — in the form of the packet-switched network using X.25 — rapidly followed, and provided another fruitful field for the application of queueing theory, as well as a whole new type of traffic to be measured, characterised and modelled. First performance studies were low level, but the intimate study of LAP-D soon gave way to X.25 above it; and there began what is now common practice — the migration of studies upwards, progressing to higher-level studies as the performance of each layer in turn is established. The lineal descendant of this is (for instance) a present-day hierarchy such as SDH — ATM — IP — TCP — FTP, where just sufficient detail is retained of one level for it to contribute to the next. At this stage, too, multi-bit-rate systems made their first real appearance, and the curious effects started to be found which are associated with the co-existence of different traffics.

As physical networks evolved to make the detailed low-level link-by-link data-checking offered by X.25 a burden rather than an advantage, data protocols simplified and evolved too; and performance engineers had to acquire skills in a wide range of bearer technologies, in particular frame relay and the slowly emerging ATM, while taking in as a minor addition the details of SMDS and other transient contenders such as the Cambridge or token-passing rings and the Orwell protocol. The co-existent quality of service (QoS) levels on these opened up new areas for study, ATM classes of service grew richer and steadily more complex, and connection acceptance controls were devised. Academic enthusiasm for this new field was considerable, and for many years there was major cross-fertilisation between the academic and engineering communities — an example of this being the effective bandwidth methodology for the VBR service class of ATM, a novel and elegant analytic technique developed for usefully characterising the traffic and allowing a sensible dimensioning procedure. For the ABR service class, however, based on the frame relay Foresight^{TM} mechanism, the complexities are such that detailed simulation remains the only real method of tackling the problem with confidence — a situation shared by the TCP protocol today.

Another new area which presented novel problems that required resolution was that of cellular mobile radio. Both the signalling protocols, with their distributed queueing mechanisms, and the dynamic channel assignment methodology, for optimising radio spectrum allocation, provided new and stimulating challenges. The latter in particular, which covered the 1970s with idealised hexagonal-cell arrays, rapidly evolved, once real cellular networks started to be rolled out, into the use of very large and sophisticated radio-planning and simulation packages taking into account radio coverage on a very detailed geographical level — complete with databases including individual houses and trees, as well as the actual topography.

About this time, network signalling arose as a major issue in its own right. While there is nothing particularly difficult or involved about the performance problems of signalling, nonetheless its separation from the transmission network, its individual treatment and processing, its complexity and importance for the operation of the network, and above all its information-bearing role in the intelligent network, all combined to establish it as a subject of importance and one where performance engineers play a crucial role.

1.4   Hardy Perennials

An area which has always been of importance, and which is likely to remain so, is that of network design and optimisation. This includes a variety of techniques and approaches, and covers a vast field — from the global multiservice network of an international carrier, down to the single-company network dominated by tariffs and equipment constraints. Performance is not restricted to merely the designing of a least-cost network, or the dimensioning of one to carry target traffic levels; but covers the whole gamut of requirements, from the identification of preferred data technologies (whether frame relay, X.25, ATM, IP, or SMDS) and resilience requirements, through physical design and routing, to a variety of real-time performance requirements covering all aspects of establishing sessions and QoS.

Associated with this is the thorny area of forecasting. No forecasting is ever accurate — but that does not reduce the drive to try to make it so. A sporadic stream of 'improved' methodologies has emerged over the years, but few of these have stayed the course since their complexities (often significant) have seldom resulted in any real, worthwhile gain. Consequently performance reports now invariably incorporate allowances and studies of the effects of load variability — which makes them apparently less definitive, but in reality more useful and realistic.

Another mathematical distraction has been supplied by 'catastrophe theory'. There are a number of areas where an ill-conditioned network may, for some operating region, exist in two different states — typically one where congestion is minimal, and one where it is very significant — and over time may flip between these two states at random. Obviously this is highly undesirable, and so the study of such bistable systems is not only interesting academically but also of major practical significance. Examples where this may happen include:

- many distributed-queueing systems (such as Ethernet, or some mobile radio signalling channels);

- some TCP-IP network scenarios when network round-trip times are perilously close to back-off intervals;

- even classical circuit-switched networks, when the provision of alternative routes through automatic adaptive routing (AAR) can at times lead to the situation where most connected calls are carried by circuitous routes and so result in widespread congestion.

Study of these, and the classification of the flips in terms of the limited number of possible mathematical catastrophes, is interesting but essentially sterile: it is only a knowledge of the actual dynamics of the processes (see, for example, Ackerley [1]) that gives rise to useful predictions and results. What has come out of them — through more conventional approaches — is the knowledge of how to stabilise networks against these phenomena, and what controls to apply.

1.5   The Current Situation

The modern era of performance engineering has little overt similarity to the classical studies done 30 years ago. At present, there are perhaps four areas of particular interest.

- Traffic characterisation

  The first area is that of traffic and applications behaviour and categorisation. This is studied in more detail in Chapter 2, and we will say no more about it here except to observe that in order to predict the real behaviour of a system we obviously need a reliable knowledge of its drivers, i.e. of the impressed traffic. At application level, the subject is in its infancy; but advanced techniques are now available to describe this at basic transport level, such as effective bandwidth, self-similarity, or wavelet techniques. While all of these have an obvious allure, none provides the performance engineer with a really useful tool, because they tend to be descriptive rather than predictive. They have, however, resulted in a very large corpus of academic literature, as well as prognostications of performance disasters and the need to rewrite all existing models. It is reassuring to record that network disaster has yet to put in an appearance, while the need to revisit the detail of existing theory has been obviated so far by the rather coarse-grained nature of most existing service-quality measures.

- Network control

  The second area is that of network control. With the varied use of networks today, a whole range of mechanisms must be deployed in order to ensure that they run smoothly; and that neither general overloads, focused events, nor any fault conditions can result in uncontrollable degradation. The foremost of these mechanisms is probably in signalling and call-control (for more information on these topics, see Chapters 7 and 8). Another is that of protecting one service on the network from another that is carried over the same bearers — whether on the voice network, over ATM, or over QoS-enabled IP (as exemplified in Chapter 3).

- Data services

  The third area is the whole arena of data services. ATM is now so well understood, and so well-performing in comparison to much that runs over it, that its performance importance is diminishing (exactly as that of LAP-D diminished as attention moved to the higher layers). IP on the other hand provides a whole range of control facilities and traffic interactions that are little understood, and a vast array of parameters that need to be set by operator or customer (see, for example, Chapter 4). In addition, it is becoming ever more important; its performance in absolute terms is (by classical carrier standards) very poor, and it is being re-engineered to provide separation, QoS, and security. There can be little surprise that this is a prime focus of much performance effort today.

- Management systems

  Finally, there is an increasing emphasis upon management systems, the back-office processes which are at least as important to a profitable business as the network QoS itself. These range in scope from traffic and network monitoring and fault reporting, to billing and customer service systems, and their remit is more typically the huge distributed database or computer system than the communications network itself. Where once the fault-lines within teletraffic were between switches and networks, circuit-switched and packet, or old technology and new, now those of performance lie more between networks and systems, or prediction and reporting.

The increasing realisation within the industry that an integrated approach is necessary, covering not only the whole hardware platform life cycle, but also the entire range of related process issues which determine every customer's experience, has resulted in a huge broadening of the scope of performance work. There has never before been such diversity or vitality in the field.

1.6   Where Now?

What then is the likely future of performance engineering?

Over the past 30 years, performance engineering has evolved its tools and techniques and moved flexibly to fill the current demands upon it. It is difficult to imagine a future in which there is no limitation in any part of the network's capacity; nor one where there is no economic pressure to reduce costs by refraining from overprovision. Certainly it does not reflect existing business conditions. We can be confident that there is indeed some future for performance engineering; but what are the major issues likely to be?

These will obviously be determined by the mixture of technologies and applications; but we can be reasonably confident that they will include:

- the protection of one service from another on common systems;

- the integration of services, or the carrying of one over another (the obvious example is voice or video over IP);

- how to accommodate, without undue loss of QoS, the short-term bursts of demand that arise naturally in data systems;

- control.

As networks get larger and more complex, and applications more aggressive and erratic, control in all its forms becomes more important. We should expect not only connection acceptance control at several levels, but also policing, shaping, reservation, priority, discard, and many other techniques. Evolution is characterised by the ongoing battle between predator and prey: as the network becomes more open, and has more intelligence, it becomes ever more tempting to predators, and ever more susceptible to wholly innocent but undesirable traffic effects. Performance engineering will certainly remain essential both to size systems initially, and to tune the controls to keep them stable.

Finally, we can expect the techniques used to be not dissimilar to those of today — which in turn are the direct descendants of those of the past. Though systems may get richer, there is only a limited extent to which our mathematics and understanding can keep detailed track of them. As applications become more complex, we can characterise them less precisely and less reliably, and this in turn limits the detail to which we can model the overall response of the network, however good our representation of the system may be. But for a particular application the situation is different; and the performance characteristics of these are likely to become increasingly critical and contentious — as is evidenced by the effort that now goes into monitoring these. While therefore we may in some sense say that studies of network performance have now reached their maximum possible complexity, those of applications may be only beginning. Evolution is by no means yet at an end!

Reference

1   Ackerley, R. G.: '*Hysteresis-type behaviour in networks with extensive overflow*', BT Technol J, **5**(4), pp 42-50 (October 1987).

2

TRAFFIC CHARACTERISATION AND MODELLING

N W Macfadyen

2.1　　　　Introduction

'Telecommunications traffic' is a phrase used to describe all the variety and complexity of the usage of a network. It is the comings and goings of demand, in response to user behaviour and to the network's reaction. Its detailed specification is a means to an end — merely the first step in the evaluation of the quantities that are meaningful and useful to the network engineer, and therefore the type and complexity of its characterisation reflects the richness of its array of uses.

Time-honoured representations of network traffic by pure-chance streams, which have worked well for many years, are now insufficient. The new data networks display a complexity of behaviour which requires a corresponding richness of input, and this has required the development of a multitude of traffic descriptions. These include not only extensions of classical models, such as multi-bit-rate sources, but also such very different areas as the effective bandwidth models now in widespread use for ATM modelling, self-similar or fractal traffics, and the complex self-coupled and adaptive behaviour of TCP/IP traffic streams.

At the same time, the output required from modelling has increased significantly. No longer is a simple figure adequate, predicting overall mean delay or probability of call blocking: instead, entire flows or streams of demands now have to be treated as a whole, and new types of analysis and statistics are required which relate to the behaviour across packets but within a stream, such as delay variation.

The uses of real importance are those which give direct network-related predictions, whether of capacity required, achievable congestion or delays, or control mechanisms. These uses drive the entire modelling process — theory is justified by its application. This essential pragmatism must be contrasted with the more academic view that sees such problems as of interest in their own right, and the field as a source of fascinating new theoretical challenges, but where the actual application of the results to problems of relevance is of only minor concern.

The practising performance engineer is familiar with all these techniques. The choice of traffic model to employ in any situation is one which has to be made with care, balancing the relative importance and advantages of all the factors concerned, including both the characteristics of the demand-stream and the sensitivities of the network. An incorrect choice can, as much as an invalid set of assumptions or an incorrect system description, lead to performance predictions and capacity requirements which are seriously wrong, and hence either to endemic poor performance or to inflated network costs.

This chapter will distinguish loosely between characterisation and modelling. The first of these implies the purely phenomenological description of traffic, i.e. the analysis of measurement data, and the production of high-level statistical descriptions. Modelling, by contrast, covers the detailed low-level production of probabilistic models of traffic that can actually be used to make predictions about network and system behaviour. The two perspectives are yoked together by the obvious requirement of compatibility — a high-level profile derived from the low-level model must fit the experimentally observed characterisation. The chapter moves steadily from aspects which are pure modelling (section 2.2) to those which are pure characterisation — the self-similarity discussion of section 2.7.

2.2   Traffic Offered

We start by emphasising the fundamental distinction between the traffic that is actually *carried* on the network, and the traffic *offered* — the underlying demand. By definition, in a voice network this latter is the traffic that would have been carried on a network of infinite capacity; for data traffic, however, this definition becomes blurred because of the interaction between user behaviour and network response, and defining terms becomes a crucial part of the performance engineer's task.

In principle, we can deduce the offered traffic from a knowledge of the carried traffic and of the relation between offered traffic level and congestion. In practice, however, this relationship is so nonlinear, and the measurement variability is so large, that it is of little use even in the simplest cases. Figure 2.1 shows two hours of (simulated) minute-by-minute behaviour of a typical telephony traffic stream, of mean 60 Erlangs, offered to a circuit-group of size 80 so that congestion is negligible. Figure 2.2, on the other hand, shows what this would be if offered to a group of size 65 trunks. It is evident what difficulties are presented in trying to work back from this to the earlier uncongested example (Fig 2.1); indeed, even recognising that this is a carried (smoothed) stream is nontrivial.

Since the goal of performance studies is to predict the effects that the network has upon the traffic, it is obviously essential to start with a model of the unaltered traffic itself. If we offer, for instance, a policed stream, we cannot be surprised when we see little further policing. For that reason, measurements which show network-related effects are of little real use, although they may be highly valuable as validations of modelling predictions or as input to subsequent stages of a network. It is chiefly for this reason that it is so important to create reliable models of the way that fresh traffic behaves.

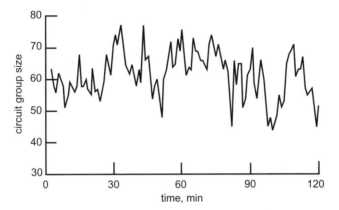

Fig 2.1   Uncongested voice traffic.

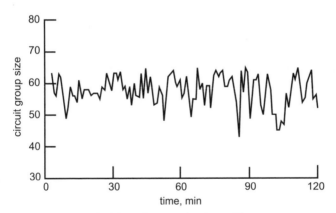

Fig 2.2   Congested voice traffic.

The fact that traffic is 'stationary' in a statistical sense does not mean that it looks smooth and well-behaved, but rather that there will be no significant changes observed if we look over a different period, provided we keep the resolution constant. In practice, it is frequently (indeed, generally) the case that the overall mean level of demand changes gradually as time progresses. Identifying and separating this out from random variation at a fixed mean is nontrivial, and requires data over a period which is long in comparison to the typical time-scale of the random variation itself. Figure 2.1 is typical: despite looking very unpleasant, this is actually perfectly well-behaved statistically. It is of course precisely because real traffic has such variability that we need mathematical tools and techniques to treat it meaningfully and consistently. Similar problems arise with data networks, due once again to the higher-layer protocols (which this time are software rather than liveware). We shall return to these in due course.

2.3 Macroscopic Variation

The most important and natural classification of offered traffic behaviour to be made is perhaps between macroscopic and microscopic variability, which are of relevance to system planners and performance engineers respectively. Broadly speaking, variation is macroscopic if it is over a lengthy period, such as a day, and microscopic if it is rapid. More precisely, variation in any parameter of a traffic stream is macroscopic if it is slow when compared with all relevant system time-scales, so that the system can for all purposes be treated as being in quasi-equilibrium at all times (the adiabatic approximation). It is microscopic if that is not the case, but if the variability is an essential part of any analytic treatment.

This distinction is necessarily pragmatic. On the one hand, no analytic treatment which ignores time-dependence can be totally valid except in the limit of infinitely slow variation; and on the other, changing the system and application-set involved can alter the classification. An example of this would be one where the view of system performance shifted from a one-hour or one-day perspective, to a longer-term one which included the network planning and provisioning processes.

Although financial modelling tends to focus upon high-level measures such as paid minutes per annum, for most network-planning or business-related applications the starting-point for any calculation is that of the mean traffic level. This mean has to be taken over some fixed period, which, traditionally in voice networks, has generally been one hour, although in certain applications other periods are appropriate, and there is now a general move towards shorter periods for new services.

The variation of this mean over the 24-hour day, with the classical 3-hump profile of telephony traffic displaying morning, afternoon and evening busy periods, is widely recognised. Newer systems, however, frequently depart from this simple picture. In the PSTN as a whole, the influx of uncharged dial-IP traffic now means that the quiet period in the night is beginning to fill; and many businesses have a profile where 'normal' day-time customer-facing or operational activity is supplemented by daily out-of-hours network data transfers, and often also by weekly or monthly company-wide data back-up operations where the apparent demand for bandwidth far exceeds what is needed during the regular morning busy-hour. Businesses are often actually unaware of their own demand profile — to acquire such an understanding requires an awareness not only of their business processes, but also of their co-operative nature and the networking implications, which is over and above their normal business knowledge. At the same time, much useful information is often obtained if a business' traffic stream can be broken down further. For that reason, decompositions by protocol or by application are often required. As an instance of this, traffic over an IP network may be categorised according to the higher-layer protocol that runs above the network-level IP itself, so that a breakdown between TCP (the majority), UDP, ICMP, etc, becomes essential. Similarly, the applications breakdown between HTTP, FTP, SMTP, DNS, etc, may be of great interest.

Associated with the systematic variation of the mean traffic over the day, the week, or longer periods is its natural variability at a fixed time. There is (for instance) a random variation in the busy-hour traffic level on each day, which is not ascribable to systematic variation but needs to be taken into account separately. Studies have shown that there is indeed a co-operative effect here, due to common social factors which affect many people simultaneously. A pragmatic, useful model of this can be created by representing the variability of the mean traffic on any day (after removal of seasonal weekday effects, which are frequently product-specific) by:

$$\mathrm{var}(a) = \kappa a + \lambda a^2$$

We stress that this is a model of the behaviour of the true underlying mean from one day to another: actual observed traffic values or measurements also contain contributions from random fluctuations about the instantaneous true mean value, which result in an increase in the first constant above. The second term here is rather small, and only becomes significant for larger parcels of traffic.

Ignoring this source of variation can result in networks which are significantly under-dimensioned. This has in fact long been recognised; for many years the Bell System practice took variability explicitly into account, while the BT approach guards against variation through the use of overload criteria, i.e. the specification of grade-of-service targets not merely at the assessed 'normal' load, but at 10% and 20% uplifts too. Other carriers have used these or different methods, but common to all of them has been the acceptance that isolated outlying days may have to be excluded from the normal processes because of their atypical nature. The classical US instance here is Mother's Day; in the UK this is not generally particularly significant, but other days are recognised as exceptional instead. There are of course also unanticipated major surges which occur for reasons like bad weather or catastrophes, and these are usually either excluded from the reckoning too, or covered by longstop positions designed more to preserve network integrity than grade of service.

2.4   Classical Microscopic Theory

The microscopic domain is that of the performance engineer. Studies here are more sophisticated, and targeted at detailed network behaviour — consequently, a more comprehensive traffic description is required.

The usefulness of any such description depends on a number of factors. Most fundamentally, it should be mathematically tractable and be able to yield useful results. In addition, it should also be:

- stable — a given stream must not have wildly fluctuating parameters from one day to another;

- parsimonious — only a small number of parameters should be necessary;

- comprehensible — the significance of the parameters should be easily understood;

- aggregatable — the parameters of the superposition of two streams should bear a simple relation to those of its components;

- scalable — natural growth in traffic should not result in complex changes in parameters.

Unless it meets these requirements, it is unlikely to become an accepted part of a carrier's armoury of operational techniques.

In the fullest detail, a complete description of a stream can always be given by the probabilistic description of the arrival process of each demand and of the work which it brings. In principle, everything can be dependent upon everything before it, and we require a multidimensional probability distribution function of the form:

$$\Pr\{T_1 < t_1 \,\&\, T_2 < t_2 \,\&\, T_3 < t_3 \,\& \ldots$$
$$\&\, S_1 < s_1 \,\&\, S_2 < s_2 \,\&\, S_3 < s_3 \,\& \ldots\}$$

where T_j is the arrival-time of the jth demand and S_j is the work that it brings (i.e. its service-time requirement). This is, however, totally intractable; it is fortunate therefore that generally (but not alas invariably) it simplifies dramatically, with the inter-arrival time between two demands being independent of anything beforehand; in technical terms, arrivals form a renewal process.

A second assumption widely made is that successive service-times are independent too. When this is the case, the traffic description simplifies dramatically, and we can write the above expression as a product of terms of the form:

$$\Pr\{T_j - T_{j-1} < t_j\} \cdot \Pr\{S_j < s_j\}$$

The most basic situation is when the arrivals have no dependence or correlation and are negative exponential — that is, $\Pr\{T_j - T_j - 1 < t\} = 1 - \exp(-kt)$. They are said to form a Poisson process. When the $\{S_j\}$ too have a negative-exponential distribution (negexp for short), the traffic stream itself is said to be Poissonian or pure chance.

These two assumptions cover between them such a large proportion of the cases of interest, and are so powerful, that they underlie almost all performance studies. We shall consider some particular instances below.

2.4.1   Arrival Processes

Actually measuring and categorising arrival processes is extremely difficult. Among the many reasons for this are the following:

- the great degree of random variability involved implies lengthy measurement — on the other hand, traffic is seldom really stationary for long enough to complete this;

- it is by no means always clear how to fit a hypothesised process to the data;

- broadly speaking, the more adjustable parameters we have at our disposal, the more precise a fit can be made — there is, however, a distinct absence of satisfactory theory to help us decide either if a fit is adequate, or even indeed if the increased number of parameters has given us any statistically significant improvement at all.

In all cases, the crucial question that determines the sophistication of the model to be adopted is simply: 'What type of statistical test does the system itself perform?' More precisely, what type of behaviour is of relevance in the studies to be conducted? On the one hand, real patterns in the arrival process which occur over time-scales much longer than the notional system relaxation time are of no concern; on the other, nor are those which are so rapid that they vanish within a time-scale short compared to the system reaction-time (such as the time taken to fill or empty a queue buffer). It is possible [1] to analyse sequences of inter-arrival times, and, by sophisticated comparisons on their correlograms, to identify strong structural patterns evidencing repeat attempts or prior shaping in the stream, even when all statistical tests not based on correlations lead to a conclusion of perfect randomness; but to use a traffic model which took this into account would generally be totally unnecessary. If an effect can only be deduced by sophisticated statistical tests upon a lengthy record of data, it is most unlikely to be of significant relevance to a system's performance.

The archetypal and simplest application of traffic theory is that of classical voice traffic over the PSTN. Apart from special circumstances which we shall consider shortly, the arrival process of this is negexp. This distribution is simple, and allows a lot of elegant mathematical theory to be derived. It possesses a number of remarkable properties, of which the following are examples:

- it is the maximum-entropy distribution (see Good [2]), i.e. if we have absolutely no knowledge whatsoever of the distribution apart from its mean value, information-theoretic considerations show that this is the appropriate one to use — all others have some structure built in to them;

- it is memoryless — the future evolution of the process is totally independent of its past;

- although the stream is described statistically by just two quantities — the mean inter-arrival time, and the mean service-time — all applications can be expressed using just the single derived quantity, the offered traffic,

- classical loss networks (Erlang theory) are actually characterised by an invariance property which means that the equilibrium blocking is independent of the holding-time distribution;

- the negexp distribution, together with the Gaussian, is one of the two great limiting distributions of probability theory — under wide (but not universal) conditions, assembling a great many small independent streams tends inexorably to a negexp.

This classical model therefore satisfies the high-level requirements of robustness, tractability and parsimoniousness to an unexpected extent.

2.4.1.1   Other Simple Arrival Processes

It is, however, by no means true that a negexp arrival distribution is always appropriate. An obvious example here is the regular arrival process of a stream of ATM cells within a single burst — the lengths of time between successive cells are all nominally identical, subject only to the small amount of variation introduced by accumulated network jitter. We describe the process as one of deterministic inter-arrivals, or, if N streams are multiplexed together, as the (considerably more complex) super-position of N deterministic processes with random mutual offsets.

On the other hand, a voice traffic stream which represents the overflow from a circuit-group is peaky — intuitively, the arrivals are likely to be bunches interspersed by lengthy gaps. This is a classical model which has long been known, but has until recently been archaic in the sense that developments in technology rendered it of little interest. New problems are now, however, reinstating its relevance, a particular instance being the way in which, to be economically viable, wholesale Internet service providers may be provided with high-usage routes and then overflow on to routes on which the native traffic must be protected.

As another example, traffic distributions with long-range dependence are occasionally modelled by the so-called Pareto distribution, with probability density $dF(x) = B(\mu, \nu)^{-1}x^{\mu-1} (1+x)^{-\mu-\nu}dx$ where B is the beta function and $\mu<0$. The rationale for this is that it is a simple function with a tail density like $f(x) \sim c.x^{\nu-1}$ as $x \to \infty$, and so represents the long-range correlation that is a feature of self-similar traffics.

2.4.1.2   Limited Source Models

In some instances the arrival process is constrained through the fact that there is an inherently limited number of possible sources. This is another instance of a classical model which, although not very relevant in its original context, still appears in new applications. The process is treated fairly dismissively in many textbooks, which frequently succeed in confusing rather than illuminating the reader.

Curiously enough, despite its superficial simplicity, this is perhaps the most challenging situation for the traffic modeller. The difficulty is caused by the fact that every source is by itself a significant contributor of traffic. Consequently, unlike the infinite-source case, we need to define rather carefully what its behaviour is not only under normal conditions, but also when it encounters congestion — specifically, its repeat-attempt behaviour (if it makes any) or the fate of the blocked attempt itself. The difficulty is compounded by other issues too — notably, the lack of clarity over what may be meant by 'offered traffic'.

While the actual mathematics of this model is not significantly more complex than that of the infinite-source, its impact is wholly dependent on the confidence we have in the model of source behaviour. Broadly speaking, when congestion or delays are small this becomes irrelevant. Away from that area, the details of the model we assume exert an overwhelming influence upon the results we obtain. Apparently subtle alterations in this model can have a major influence upon the behaviour of the entire system being studied; careful exegesis and setting-in-context of results is absolutely essential.

2.4.1.3   *Markov Modulated Poisson Processes*

If the arrival process at any instant is standard Poisson, but switches at random between a (finite) number of different levels, it is said to be a modulated Poisson process. In the case where that level-switching is itself a Markov chain (i.e. where its switching probabilities do not depend on the past history), the overall process is termed a Markov-modulated Poisson process (MMPP).

An MMPP over n states is described by n^2 parameters — the n arrival-rates in the several states, and the matrix of $n(n-1)$ transition probabilities between them. The progenitor of the MMPP, the interrupted Poisson process (IPP), is over just two states, conventionally termed 'on' and 'off'. Since the arrival rate in the off state is by definition zero, this has just three free parameters to describe it.

The MMPP owes its theoretical popularity to the fact that an elegant closed-form analytic solution, using a matrix-geometric formalism, can be found for many teletraffic problems where it occurs. Its practical application, however, has been distinctly limited. With a process on three states only, there are nine adjustable parameters required; and fitting these poses a generally insuperable difficulty. In occasional instances, the states and transitions arise naturally in a well-defined manner from the operation of the system concerned, and in such cases the method is (in principle) useful; but where the characteristics of a traffic stream have to be fitted, whether that stream is a real observed (that is, measured) stream or the output of some other sub-model, serious difficulties arise.

For all these reasons, the MMPP has had a rather limited application in practice, however academically satisfying it may be. It is very seldom that there is sufficient confidence, in either the traffic model or in the system response, to justify such an elaborate representation. Its simpler brother, the IPP, does, however, play a significant role in ATM modelling (see section 2.5).

2.4.1.4   Wavelet-Based Analysis

Wavelet analysis (sometimes termed multi-resolution analysis) breaks the data down into different 'frequency components'. It can be regarded as growing out of an attempt to 'localise' Fourier analysis by replacing the pure harmonic components by wavelets of finite range. While technically possible, this is open to the charge of arbitrariness — once we abandon the underlying symmetry group whose representation theory underpins classical analysis, there is no canonical choice of wavelets or natural and unique interpretation of the transform.

Unlike the parametric methods above, very little needs to be assumed about the underlying process. This latter is its underlying weakness — because little is assumed about the dynamics, little can be said of a predictive nature. Like effective bandwidth and self-similarity (which are discussed below), it is essentially a descriptive technique rather than a predictive. As a consequence, it has made little impact upon practising performance engineers. We mention it here only for completeness.

2.4.2   Service Requirements

The discussion above focused upon the arrival process. Traffic is also characterised, however, by the amount of work that it brings — in classical terms, its holding time. Changes in the statistical distribution of this may have as great an effect as changing the arrivals.

There are a number of contenders of real importance here. These include the negexp distribution we have already considered, and the deterministic (constant) distribution such as the demand from a stream of ATM cells.

Another is the distribution observed in practice in a packet network. Typically, IP packets (for instance) have a multi-modal distribution — there is a preponderance of minimum-length control packets and a large number of maximum-length packets associated with large data-transfers, those fewer in between being either of that length in their own right (uncommon) or being residuary fragments resulting from fragmentation of MTUs to fit system capabilities.

The distribution observed can be highly variable. Typically, downstream and upstream can be quite different, and there may be major changes over the course of a day — Fig 2.3 illustrates this. This is drawn from frame relay (FR) data, from a

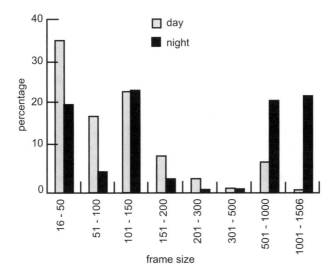

Fig 2.3   Typical frame relay frame size distributions.

single customer; it shows the frame-length distributions on a normal weekday (left-hand bars), and then over a weekend night (right-hand). They reflect the complete change in applications mix between the two periods. We note incidentally that the night-time period is actually very busy — it is by no means a low-traffic period.

The actual mean frame size levels in Fig 2.3 (in bytes) are shown in Table 2.1. The upstream/downstream asymmetry in those ADSL traffic streams which have video sources is, of course, even more extreme.

Table 2.1   Mean FR frame sizes (bytes).

	Transmit	Receive
Daytime	217	125
Night-time	523	89

2.5   ATM Traffic Modelling

ATM traffics and networks have been intensively studied over many years, and have proved a rich area for innovative mathematics. Absolutely key to their modelling is the recognition of the traffic's time-scale structure.

There are three relevant time-scales involved — the cell, the burst, and the connection scale — which differ by several orders of magnitude. If we are concerned with cell delay variation, the cell level is appropriate; for connection acceptance control (CAC) the connection (or call) level is the right time-scale; while for many applications it is the comings and goings of bursts which are relevant.

Because of the considerable differences in duration, each time-scale can be studied quasi-independently, assuming as given the behaviour below it, and ignoring any changes above it. Recognising and capitalising upon this is important not only for analytic modelling, but also for the design of simulation experiments and the confidence that can be placed in their results. Its judicious application is part of the stock-in-trade of the experienced performance engineer.

A practical modelling approach is to assume that the data arrives in variable length bursts at a known peak value P Mbit/s with both burst and silence lengths having a negative exponential distribution, giving rise to a given mean value m Mbit/s. This is a reasonable approximation for traffic aggregated from many sources and applications, although it may not be so for a source running a single application using, for example, fixed length packets.

With these and similar assumptions, the traffic stream can be adequately modelled, at least for such purposes as the derivation of traffic shaper performance, leaky-bucket policer losses, queueing buffer overflow statistics, etc. Representing sources of traffic as an IPP or MMPP allows a wide range of scenarios to be modelled. Much work has been carried out on this, and sample results can be found in the contributions to the COST 224 and COST 257 projects and their final reports [3, 4].

2.5.1   VBR — Effective Bandwidth

A specific area of ATM that is of particular interest and relevance is the variable/statistical bit-rate (VBR) service-class. This has also excited considerable mathematical attention, because constant/deterministic bit-rate (CBR) is of little interest; and available bit-rate (ABR) is both little used and extremely difficult to treat because of its inherently adaptive nature, being consequently largely studied by simulation, where every detail can be reproduced.

For the VBR class, the typical situation of interest is one where there are a number of traffic streams, each varying independently — i.e. providing varying amounts of work to the server (or transmission link) — and over very short time-scales. The question of concern is: 'To what extent is it possible to fit together these streams and serve them with a resource that is (obviously) greater than their mean but less than the sum of their peaks, in the statistical assurance that these peaks will not significantly clash?' Clearly the question is of crucial importance to every carrier that wishes to benefit from multiplexing gain.

Straightforward models of time-varying traffic are not generally applicable, because the crucial issue is that traffics are varying independently, and we have to model not only the individual streams but also their relative positioning. The description used is therefore in terms of the elegant mathematical concept of effective bandwidth (EB). This is not in its own right a traffic model, but rather a high-level statistical consequence of a traffic model.

To introduce this, let the (stochastic) amount of work brought by an arrival process in the interval (0, t) be denoted by $X(t)$. A simple way of conveniently enumerating its statistical properties is through its moment generating function $M(X)$, defined by:

$$M(X)(s) = \sum_i \frac{s^i}{i!} E(X^i)$$

For a demand stream Y which is absolutely constant, at rate (bandwidth) α, say, we should have:

$$M(Y)(s) = exp(t\, s\, \alpha)$$

and so it makes sense to introduce the effective bandwidth function $\alpha(s,t)$ of the process by:

$$exp(s\, t\, \alpha\, (s,\, t)) = M(X)(s)$$

or, explicitly:

$$\alpha(s, t) = \frac{1}{st} \ln E\{e^{sX(t)}\}$$

This effective bandwidth function $\alpha(s,t)$ has many useful mathematical properties, in particular:

- it is additive — if X and Y are two independent processes, then the effective bandwidth function of the sum process is the sum of the individual effective bandwidths;

- if the process X is one of independent increments, then $\alpha(s,t)$ is independent of t (note that this is not generally the case — the common case where X is modelled by an IPP is not covered).

A full discussion of its properties is outside the scope of this chapter. We need only note here that its usefulness lies in the fact that it is mathematically quite tractable, and we refer to Kelly [5] for an introductory account. Its real usefulness hinges upon the fact that a straightforward explicit expression (in some cases approximate, in others a bound) can be derived for the overflow rate of a server with a queueing buffer of a given finite length, and this leads directly to appropriate system dimensioning.

Turning this round, if we have a server with a fixed buffer size, then, once the allowed cell-loss rate is defined, the theory provides a well-defined value for the maximum effective bandwidth (EB) of a particular traffic stream. It is this which is generally referred to in practice as the EB of the stream, i.e. it is defined by $EB = c/N_{max}$ where c is the link rate, and N_{max} is given by:

$$N_{max} = Max\ \{N : Pr\{Q \geq buffersize\} \leq \varepsilon\}$$

where ε is the given allowed cell-loss probability.

We illustrate this below, with a pre-encoded MPEG video-stream such as would be used by a large-scale on-demand video library service. This has been drawn from the set of such sequences which have been made publicly available by Rose [6] from the University of Wuerzburg. Each sequence consists of 40 000 frames, of resolution 384 × 288 at 25 Hz, and therefore represents about half an hour of video. The sequence illustrated is from the film Goldfinger.

Figure 2.4 shows the traffic profile over the monitoring period, and reveals sustained periods of higher activity. (The sustainable cell rate (SCR) used here has been chosen so that the maximum buffer required is one second. The EB has been calculated from the data using a large-deviation approximation, so that the detailed procedure of the study is rather the inverse of our exposition above.) Figure 2.5 shows the associated buffer queue lengths, from which the important fact is obvious, that delay and loss are not spread evenly throughout as one might naively expect, but are of course concentrated upon particular passages.

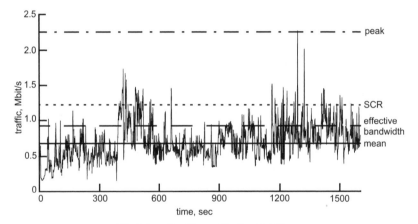

Fig 2.4   Traffic profile for *Goldfinger*.

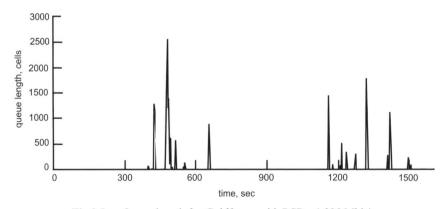

Fig 2.5   Queue length for *Goldfinger* with PCR = 1.225 Mbit/s.

Knowing precisely the effective bandwidth of all available sources individually is likely, however, to be unfeasible in practice in any situation other than pre-encoded video libraries, and so it is useful to see if there is some generic EB that could be used in the absence of further or detailed information.

Figure 2.6 illustrates the difficulty of defining such a generic EB model for video. The curves show the maximum buffer length needed (note the logarithmic scale) with a particular cell-loss rate, as a function of the SCR. As the SCR increases, less buffering is required. The different traces are drawn from those discussed above, and the feature of interest is the great variability of the traces. Films are by no means all equivalent.

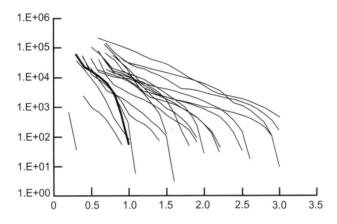

Fig 2.6 Delay as a function of SCR.

It is of interest in this context to remark that the Star Wars excerpt, which for no very good reason forms the basis of a high proportion of studies, is rather atypical. It is shown as the thicker line in Fig 2.6.

The huge variability in the parameters of the sources severely limits any attempt to ascribe a common set of parameters to films generally. Moreover, as Fig 2.5 shows, the losses and delays which will occur in practice are not spread evenly, but are frequently concentrated at fast-moving or spectacular moments of the film — which may be just those where service degradation has most impact upon the viewer. Even with a detailed knowledge of the statistics of the source, therefore, the approach of treating the film as a whole may suffer from unacceptable defects in practice.

For other applications, however, EB can provide a useful statistic with which to characterise VBR behaviour succinctly. It has also been applied in identifying optimal strategies in different tariff scenarios.

Finally, it should be remarked that EB is one of the few techniques available which can be usefully extended to pure self-similar traffic (see section 2.7).

2.6   Complex Arrival Processes — IP Traffic

The traffic processes associated with IP networks show a new variety of richness and complexity. IP itself is not especially interesting from a traffic viewpoint — it is merely another packet data protocol, which can be studied by the same techniques that have long been in use for LAPD and X.25 — but its associated suite of protocols is another matter. Furthermore, the characteristics of the network and its traffic are rather different from those of classical systems.

2.6.1   Observed Macroscopic Behaviour

We note first that there are a number of quite different perspectives on Internet traffic, ranging from access through core to application. Much traffic originates from residential users who access the network through dial-up PSTN links terminating at a modem (typically at a nominal 56 kbit/s, although compression allows the apparent data-rate to exceed that figure slightly). At this level, we can categorise it by a variety of standard PSTN measures such as profile over the week, holding-time, etc. Apart from the profile, the factors of interest include the variability of mean holding-time, and the aggressive re-dialling behaviour upon congestion. Autodiallers are generally configured by default to some ten repeats, and these naturally give IP traffic an advantage over the native voice, unless precautions are taken to protect the latter.

Figure 2.7 shows a typical profile of holding times for calls starting at a particular time. This is strongly affected by the tariff structure in operation, with a spectacular increase at weekends. The daily spikes are caused by the interaction of the tariff-period boundaries with a mechanism for disconnecting calls which have lasted more than a certain duration.

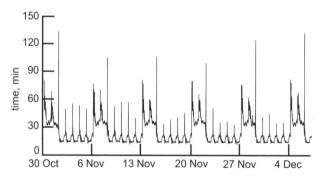

Fig 2.7   Holding-times of IP-dial traffic (for details, see Chapter 3).

Figure 2.8 shows the mean session rate (that is, the overall average data-transfer rate over an entire user session) plotted against the number of concurrent dial-up users — which of course is related to the network congestion. There are several distinct groups of user identifiable, who can be categorised independently and fall into different broad patterns of usage. Although there is a significant scatter, it is clear that the typical user tends to a well-defined maximum overall average data rate requirement. The evidence is strongly against the alternative hypothesis, that faster transfer results in shorter sessions. It should come as little surprise that the behaviour of broadband residential customers (those with ADSL, providing downstream rates of up to 512 kbit/s) is very different.

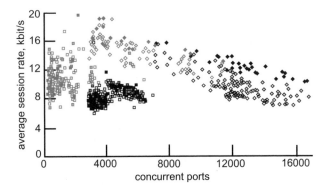

Fig 2.8   Session rates and occupancy.

2.6.2   Network Traffic

Traffic on LANs has been shown in numerous studies (see, for example, Leland et al [7]) to be self-similar; in large networks, however, it has been found [8] that the overall arrival rate at nodes within the core of the network is remarkably Poisson. The distribution of packet lengths, of course, is nothing like negexp. Figure 2.3 illustrated a typical distribution on a downstream link away from a source of data. That in the opposite direction is frequently very different, consisting of acknowledgements and relatively short request packets, and with few, if any, long ones.

The performance of TCP networks is highly complex, with many interlocking factors — an introduction to it can be found in Chapter 9. The factor that is of overriding importance to the performance engineer, however, is that an individual traffic stream has to be treated as an entity. Whereas for classical applications it is the fate of a single demand that is of interest, for TCP we are concerned with the mean transfer rate achieved from the entire sequence of packet demands. While this is of course true for all data systems, only for frame relay and ATM ABR is there the complex feedback between network and source behaviour that is such a feature of TCP.

Figure 2.9 shows a typical set of network traffic measurements taken at 1-min intervals; for interest, we have chosen one which clearly shows the effect of a pair of

SDH re-routes. The curves show the link utilisation (the uppermost curve, on the left-hand axis); and the mean, maximum, and next-highest values of round-trip times measured. The great disparity between the stable mean and the highly erratic peak delays is evident.

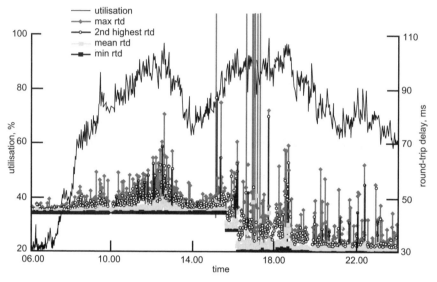

Fig 2.9   Internet traffic.

Finally, an excellent illustration of the observed stochastic delays as a function of utilisation can be found in Chapter 9.

A second major area of importance, besides sheer throughput, is that of packet jitter — the variation in timings between packet arrivals at their destination. While this is not of much concern for TCP, it is of overriding importance (more so, indeed, than is actual loss) for the real-time services carried by the UDP protocol. These must be buffered before playout, and high variability will require lengthy buffers and increase overall latency. A latency which is acceptable for a service such as video broadcasting will be quite unacceptable for interactive voice.

Modelling of jitter is complex, and an area which is still rapidly evolving. Simple queueing models applied at each node of the network independently yield results which are fully compatible with network simulations, and in excellent agreement with measurements taken of the static (equilibrium) network behaviour. Experimental observation, however, gives results which are rather different. Packet delays can be highly erratic, with bursts an order of magnitude greater than the background random jitter, which is in agreement with theory.

Figure 2.10 shows a sequence of round-trip times, on a minute-by-minute basis, to demonstrate the type of phenomenon that occurs. The distribution shows a stable and well-behaved variation at a base level, with some isolated points that are

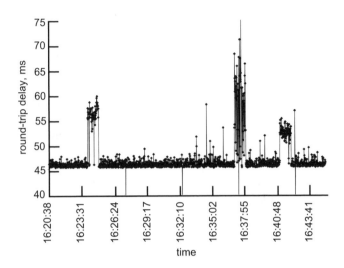

Fig 2.10   Jitter events.

(perhaps) believable, and three intervals where clearly something else has occurred. The four downward spikes are artefacts.

The effects here show quite clearly the difficulties which must be overcome in giving a useful statistical description, let alone an analytic calculation, of jitter. As Fig 2.10 shows, jitter is quite small (in the region of 0-3 ms) for a large part of the time, but with the occasional period where it suddenly rises to some 25 ms. A model which ignores the latter periods would be very seriously misleading overall; on the other hand, constructing a formal statistical distribution which includes all the data, but does not recognise the obvious correlation structure, would be almost equally unhelpful.

The cause of these apparent anomalies is of course the behaviour of the router itself, which has significant activities to perform which impact upon the basic packet-processing. Examples of this include routing updates, housekeeping, OSS activities and MIB requests, etc. It should be appreciated that the router is itself a switch, and like all large switches has a processing burden which limits its performance. A detailed understanding of this behaviour and of its impact upon different types of packet processing is still being developed; the differences between different router types and operating systems are clearly very considerable.

2.6.3   Modelling IP Traffic

An IP network's traffic is therefore a superposition of many streams, based on many protocols. The first set of these is all those associated with management — which includes routing updates. These are quasi-periodic processes, that run at intervals and therefore present the network with periodic tranches of traffic. Their overall effect is, it is hoped, generally small, once they have been taken into account from the point of overall traffic volumes, because they are (or should be) dwarfed by the flows of 'real' data.

The second system of protocols, however, causes real difficulty. While the real-time UDP is no problem, the layer-4 transmission control protocol that is commonly used for file-transfer applications — including Web browsing — is another matter.

TCP has been specifically designed and optimised as an adaptive protocol. It is a sliding window mechanism which uses regular acknowledgement packets to sense network congestion, whether through time outs or through packet loss, and then adjusts its sending rate and window size in response. Consequently the performance of a TCP traffic stream is inextricably bound up with that of the entire network in a complex way — the case is very similar to that of the ABR traffic class of ATM.

Characterising TCP traffic is therefore nontrivial. A single file transfer will indeed progressively speed up, until it is limited by network bandwidths or by packet loss rate. The only meaningful description of offered traffic which is independent of network response is therefore in terms of the overall amount of data transfer requested by each fresh arriving demand.

When this is placed further in context, however, matters become still more involved. A typical business scenario would be one where a fixed number of files of data are to be transferred — a faster transfer-rate means simply that the task is finished sooner. For a Web-browsing application, however, it seems that in many cases a more realistic scenario is that the user will browse for a fixed period of time. Speeding up the Web, then, has the consequence that a larger amount of data is requested and transferred.

Typically, performance studies look at the detailed structure of a single flow, against the background of all the other flows in the network. The analysis of the former is discussed in Chapter 4, and with even the simplest uncoupled representation of the background is highly complex; the observed richness and adaptive nature of that whole background traffic too makes this area a supreme challenge for performance modelling.

2.7   Self-Similarity

A development of recent years has been the identification of the self-similar nature of many traffic streams. This was first found in studies of internal Ethernet traffic on the internal Bellcore LAN [8]; and has since been found in many other situations. It seems to be a widespread feature of all types of data traffic.

Phenomenologically, self-similarity is the expression of the fact that some stochastic processes seem to look exactly the same over a wide range of time-scales. Furthermore, taking averages in order to estimate the true underlying mean does not result in a well-behaved estimate that gets increasingly tight as the sample size increases. This is explained formally by observing that the data is exhibiting long-range dependence and correlation; and so parts of the series, though separated by a long distance, are still yoked together by some mechanism. Obviously this is a matter of degree, and so we might expect that many different series exist which show different degrees of such self-similarity.

In practice, this is expressed by the so-called Hurst parameter. The range of this is over [0.5, 1) — a totally memoryless process such as a classical traffic process takes the value 0.5, while 1.0 would correspond to a process which was exactly the same in all respects, on whatever resolution we viewed it. Values outside this range are unphysical. Figure 2.11 illustrates this, for (artificially generated) processes with $H = 0.5$, 0.7 and 0.95 respectively; and the time-scale increasing by a factor of 16 between each set of graphs. The vertical scale remains constant.

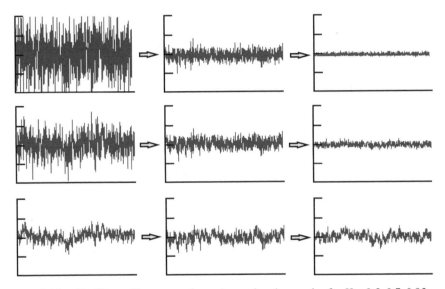

Fig 2.11   Traffic profiles averaged over increasing time-scales for $H = 0.5, 0.7, 0.95$.

It should be clearly realised, however, that self-similarity is not a traffic model; any more than a statistical statement that two random variables are not independent and not Gaussian constitutes a model of what their distribution actually is. Like the effective bandwidth discussed in the last section, it is a rather high-level statistical statement; but because it is statistical rather than probabilistic — that is, a summary of what has happened rather than a useful prediction of what will — its use is very limited.

We start with some definitions. Let $X = \{X_j \; j=1...\infty \}$ be a stochastic process. Then X is short-range dependent (SRD), if asymptotically the long-range auto-correlations $r(k)$ decay exponentially fast with k:

$$r(k) \approx A e^{-\alpha k} \qquad k \to \infty$$

In contrast, it is called long-range dependent if the autocorrelations $r(k)$ decay like:

$$r(k) \approx A e^{-\beta ln(k)}$$

so that:

$$\sum r(k) = \infty$$

Let $X^{(m)}$ be a new process obtained by averaging X over non-overlapping blocks of size m (equivalent to measuring traffic at a coarser timescale). X is then termed exactly self-similar, with Hurst parameter $H = 1-\beta/2$, if:

$$X \equiv m^{-H} X^{(m)}$$

i.e. if $X(m)$ is statistically identical to X within a scale factor of m^{-H}. This is referred to as the scaling property of self-similar traffic. This requirement of being statistically identical is, in practice, rather strong (and impossible to verify). A weaker definition is that X is second-order self-similar with Hurst parameter $H = 1-\beta/2$ if, for all $m = 1,2,...$:

$$\mathrm{var}\,[X^{(m)}] = \sigma^2 m^{-\beta}$$

$$r^{(m)}(k) = r(k) \qquad k \geq 0$$

i.e. if the processes $X^{(m)}$ and $m^H X^{(m)}$ have the same second order characteristics for any $m > 0$.

For a Markov process (e.g. negexp), with $H = 0.5$, the variance of the mean decreases with the sample size as:

$$\mathrm{var}\,[X^{(m)}] \approx a_1 m^{-1}$$

For a self-similar process, however, the variance of the mean decays more slowly than this. Establishing the asymptotic rate of decay, either directly by a variance/ time plot or via the rescaled adjusted range statistic, forms the general basis for estimating H.

The classical example of self-similarity is the process of packet arrivals on a link at different time-scales. Bellcore WAN Ethernet traffic was found to be character-ised by a Hurst parameter which, over the four-year study period between August 1989 and February 1992, varied between 0.80 and 0.95, roughly increasing with time as the host-to-host traffic decreased and the router-to-router traffic increased. It was also noted that it was higher during the busy hours, contradicting the general notion of Poisson modelling that traffic smoothes out as the number of sources increases.

2.7.1   SMDS Traffic

Measurements of traffic on the BT SMDS data service have shown a marked vari-ation between individual customers, with some being very bursty even after averaging over 15-min intervals. The distribution of the Hurst parameter *H* between customers with access class between 4 and 10 Mbit/s in one particular series of measurements, and as calculated from the rescaled adjusted range plot, is shown in Fig 2.12.

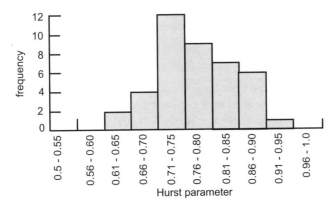

Fig 2.12   Distribution of Hurst parameter (SMDS customers).

With mean values ranging from a few kbit/s to 5 Mbit/s, and the peak/mean ratio for these few customers ranging from 1.5 to 15, the range of values emphasises the difficulty in describing an 'average' customer, even for a high-speed service such as SMDS, where the underlying management traffic and network controls are lost in the volume of traffic.

Although individual traffic sources, or even customers, cannot reasonably (that is, with stability and confidence) be modelled at the network layer, it is, however, possible to model the aggregate traffic. Figure 2.13 shows the aggregate traffic from the customers mentioned above, taken at 10-sec intervals. This was taken in 1995; since then the burstiness (and the mean) has increased further.

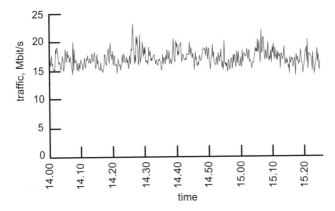

Fig 2.13   Aggregated SMDS traffic.

The aggregate traffic properties are summarised in Table 2.2.

Table 2.2   Characteristics of core SMDS traffic.

Peak 10 sec/busy hour mean	1.32
Busy hour mean/daily average	1.55
Multiplexing gain	2.5
Hurst parameter, H	0.86

It should be noted that even with such aggregation, the Hurst parameter does not reduce, illustrating the fact that aggregated self-similar traffic remains self-similar. Indeed, comparing these with Fig 2.12 shows that the Hurst parameter of the aggregated series is substantially greater than the mean of the Hurst values of its components.

2.7.2   Other Data Traffics — Frame Relay

On-line traffic measurements from frame relay customers in the UK behave similarly. Figure 2.14 highlights the importance of measuring at the appropriate time-scale. Figure 2.14(a) shows aggregated traffic from a number of sites belonging to a single customer, taken over 15 hours at 1-min intervals, while Fig 2.14(b) shows the traffic from the same customer over 35 minutes at 2-sec intervals. At the shorter time-scale the data peaks frequently at almost twice the rate of the 1-min peaks.

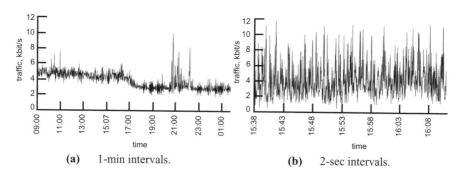

(a)   1-min intervals.   **(b)**   2-sec intervals.

Fig 2.14   Frame relay traffic profiles at different time-scales.

Figure 2.15 shows the variance/time plot used to calculate the Hurst parameter. The graph is gratifyingly linear, and suggests a figure of approximately 0.7. Clearly the choice of traffic properties, such as mean, peak and burstiness, will depend very much on the time-scale in which the user is interested.

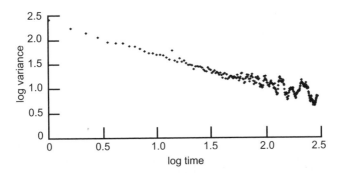

Fig 2.15 Variance/time plot.

When a single customer site is considered, however, as opposed to an aggregate, management traffic frequently has a significant effect. The danger of using the Hurst parameter alone in modelling data traffic is apparent from the first set of graphs in Fig 2.16, which show the transmit and receive traffic for a single customer site, together with the variance/time and rescaled adjusted range plots for calculating the Hurst parameter.

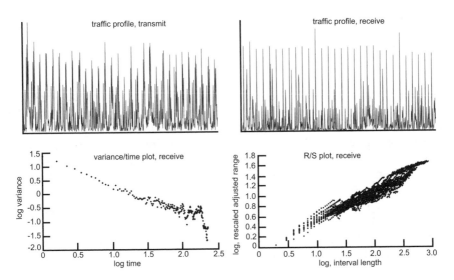

Fig 2.16 Plots from a single customer site.

The traffic here is dominated by regular management traffic, revealed as periodic peaks in the traffic profile. The Hurst parameter for these traces is approximately 0.55, which is very nearly Poisson. Using such a model by itself, however, would considerably underestimate the 2-sec peaks that occur periodically (the peaks being 10 times the mean bit rate).

Figure 2.17 shows FR traffic on a single link but to a number of destinations. The profiles belong to the same customer, but at different times of the day. The first of these has a Hurst parameter of approximately 0.54, so it is again almost Poisson, and, this time, the management traffic is of less significance in comparison with the rest.

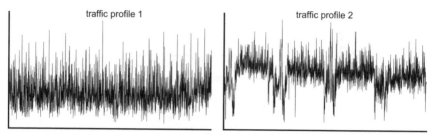

Fig 2.17   FR traffic from a single customer at different times of the day.

The second traffic stream has a Hurst parameter of approximately 0.86, which is significant. What is obvious, however, is its pronounced structure. This suggests that, although characterising it as self-similar, and describing it by the single Hurst parameter, may be possible, what is likely to be much more useful would be to ascribe a specific model such as a sequence of file-transfer activity.

Figure 2.18 is still more interesting. This represents traffic from exactly the same source, but at yet another period of time. The rescaled adjusted range plot associatedwith it is difficult to interpret — a fit by the straight line required for self-similar traffic is somehow not very plausible. Indeed by ignoring the points before the knee in the curve and basing the estimate on the latter half of the graph, the gradient suggests a Hurst parameter of 0.35, which is in the unphysical region. We show the bounding lines corresponding to $H = 0.5$ and $H = 1$.

These graphs clearly illustrate the extreme variability of a customer's traffic. It may be argued that the self-similarity should be calculated based on a longer period, averaging out some of these anomalies, or alternatively that a 2-sec resolution is inadequate. The entire interest and usefulness of self-similarity lies, however, in the fact that it applies across all time-scales, and relates the burst-scale characteristics unambiguously to those observed at typical measurement intervals. When the characteristics change (regularly) over the course of time, this destroys the notion that a single universal description is adequate for a given source; and when a specific time-scale makes itself apparent in the data, that provides the basis for constructing more specific models which incorporate better knowledge of the system.

2.7.3   Modelling Self-Similarity

The need for care in selecting a traffic model is evident. The same is true whether quasi-analytic or simulation techniques are employed; extremely lengthy simulation

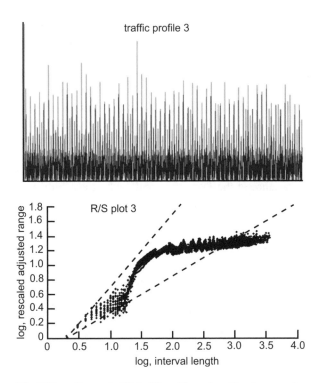

Fig 2.18 Example FR traffic with unclear Hurst parameter.

runs are required, and, while models such as fractional Gaussian noise [9] provide a good mathematical model for self-similar traffic generators, they do not take account of the observed real limitations of the theory. An alternative technique is to simulate traffic using a random midpoint displacement method [10], with the Hurst parameter as a function of the time-scale. All these techniques have their drawbacks.

An alternative approach to representing the traffic attempts is to describe it by a single heavy-tailed probability distribution, typically Pareto (see section 2.4 above). Internal measurements of LAN traffic have shown that this can be reasonably characterised by such a distribution, with $v \sim -0.7$; but the adequacy of such a representation is naturally very dependent upon the traffic streams concerned and the application for which it is intended.

2.7.4   Application — Short-Term Peaks

An area where self-similarity theory does provide useful insight is, however, the relation between short-term peaks in the traffic and the actual measurement data available. These latter are typically taken at 15-min intervals — sheer data volume and processing overhead argue against much finer resolution — and therefore do not capture the short-term structure. Self-similarity causes significant changes in the size of the short-term peaks that are inferred.

Measurements of IP over ADSL traffics have shown this to be self-similar, with a Hurst parameter of typically 0.8, and 1-sec downstream peaks exceeding 15-min measurements by a factor of perhaps 5. A simple normal approximation to the burstiness, with variance related to time-scale according to $exp(-(2H-2).ln(t))$, holds well for time-scales shorter than 15 min; but for longer periods the actual (observed) burstiness falls much faster than the model predicts. Over 10 days or so this falls to a level much the same as would be expected of Gaussian traffic with the same mean and variance at the 1-sec time-scale, whereas a truly self-similar model would expect a figure around 50 times that value!

2.7.5   Critique of Self-Similarity

Self-similar models fulfil few of the ideal requirements of a traffic descriptor which were set out in section 2.4. A hint of the lack of stability has been given above; and combining or growing streams is manifestly nontrivial. The parsimonious nature — and hence the comprehensibility — is achieved only by subsuming all sources of variation into the single Hurst parameter, which therefore has to perform multiple roles and characterise all conceivable traffic sources (thus doing none adequately); and mathematically it is totally intractable.

Much of the reason for this is the same as that for effective bandwidth not being widely useful outside one very specific context — it is a statistical description, rather than a probabilistic. Moreover, it is the very essence of exact self-similarity that the process looks the same at all time-scales: it therefore follows that there is no hope of using it to predict the future (in a probabilistic sense) in any meaningful way, since to the series a millisecond is just as far away as a year.

However, once the exact self-similarity is relaxed — typically, by confining the range of similarity to periods between the system line rate and a day — several things follow. Firstly, realistic and believable models which represent the underlying causes of the traffic become achievable; secondly, with these models analytic treatment becomes possible; and thirdly, the intrinsic mathematical interest in the technique disappears.

Finally, it should be remarked that despite predictions that the existence of self-similarity would require all performance theory to be re-written, little progress in this direction has actually been made. To a large extent, this is because of the reasons discussed above, and the intractability of any mathematical formulation; but it also reflects the fact that, empirically, systems have not collapsed, and quality of service does not seem to have been catastrophically affected.

While there have been significant changes in the approach taken by performance engineers because of this new understanding of traffic, the techniques already available have proved (in experienced hands) to be sufficiently powerful and flexible that their predictions are usefully robust. It should be observed in this context that what is sauce for the goose is also sauce for the gander: while long-range dependence means that traffics do not become smooth as they are averaged over longer durations, it also implies that they do not become as rough as expected if durations are reduced.

Ensuring that the limitations of theory and understanding are balanced, and that data requirements are appropriately adapted to this new type of traffic, allows the judicious application of the great majority of the existing heritage of performance theory in this new context.

2.8   Summary

This chapter has given an overview of some of the methods in current use for modelling and characterising traffic.

It is the emergence of new services and systems, and their increasing integration, that has driven these traffic models. While classical systems frequently are surprisingly robust to changes of traffic behaviour, new packet-data networks with QoS-aware services are not; and while a broad-brush classical model may indeed generate sufficiently accurate estimates of certain high-level quantities (for instance, call set-ups rejected due to congestion), many newly relevant service characteristics necessitate a richer model and more refined analysis. Consequently, although the classical traffic models covered briefly in section 2.4 still dominate the armoury of the operational network planner, the professional performance engineer who is concerned with quality of service, in all its aspects, has to be familiar with a much broader field.

Confidence in any performance prediction requires confidence in every component of the underlying model, and that includes of course the representation of the traffics. While generally no effort is spared in the modelling of system detail, it is all too frequently the case that the details of both the traffic demand and of the user requirements are overlooked. This is perhaps the single area where the experienced performance engineer adds most unexpected value, by identifying those features whose interaction is most critical and where significant implications can arise for the network or its users.

With newer and more integrated networks and increasingly complex systems, the potential for disastrous synergistic effects built upon apparently insignificant causes has never been so great — nor has the breadth and depth of knowledge and insight required from the performance engineer.

Without that experience and skill, the opportunities for inappropriate engineering, resulting either in overprovision or in widespread congestion (or even in both simultaneously) have never been greater. Time spent in performance analysis is time extremely well spent.

References

1   Macfadyen, N. W.: '*Statistical observation of repeat attempts in the arrival process*', Proc 9th International Teletraffic Congress, Torremolinos (1979).

2   Good, I. J.: '*Maximum entropy for hypothesis formulation*', Ann Math Stat, **34**, pp 911-934 (1963).

3   Roberts, J., Mocci, U. and Viratmo, J. T. (Eds): '*Broadband network teletraffic*', Final Report COST242 (1996).

4   Tran-Gia, P. and Vicari, N. (Eds): '*Impacts of new services on the architecture and performance of broadband networks*', Final Report COST257 (2000).

5   Kelly, F. P.: '*Notes on effective bandwidths*', in Kelly, F. P., Zachary, S. and Ziedins, I. B. (Eds): '*Stochastic networks: theory and applications*', Royal Statistical Society Lecture Notes Series, **4**, Oxford University Press, pp 141-168 (1996).

6   Rose, O.: '*Statistical properties of MPEG video traffic and their impact on traffic modeling in ATM systems*', University of Wuerzburg, Institute of Computer Science Research Report Series, Report No 101 (February 1995).

7   Leland, W., Taqqu, M., Willinger, W. and Wilson, D.: '*On the self-similarity of Ethernet traffic*', SIGCOMM'93 (1993).

8   Cao, J., Cleveland, W. S., Lin, D. and Sun, D. X.: '*Internet traffic tends towards Poisson and independent as the load increases*', in Holmes, C., Denison, D., Hansen, M., Yu, B. and Mallick, B. (Eds): '*Nonlinear Estimation and Classification*', Springer, New York (2001).

9   Norros, I.: '*A storage model with self-similar input*', Queueing Systems, **16**, pp 387-396 (1994).

10 Chen, F. H. M., Meelor, J. and Mars, P.: '*Comparisons of simulation algorithms for self-similar traffics*', Proc 13th UK Teletraffic Symposium (1996).

3

POTENTIAL INTERACTIONS BETWEEN IP-DIAL AND VOICE TRAFFIC ON THE PSTN

N W Stewart

3.1 Introduction

The PSTN has been around for a long time and its performance is well understood. The technology is mature and the issues regarding the management of the network are, for the most part, also well understood. We can see this in the excellent performance and network availability delivered by the PSTN — a record that is to be envied by some younger network technologies.

The world does not stand still, however, and the loads placed on the PSTN change, together with our expectations of what the network could deliver. Many facilities have been added to the network without disruption to the network capacity, such as the 1471 service, but other services have stressed the network such that great ingenuity is required to manage them and their impact on other traffic. A classic example of this is the televote, a modern phenomenon capable of generating enormous numbers of call attempts to a small number of destinations, over very short time-scales. Offering such services must be done without causing an adverse effect to existing services and traffic that share the network. So, as the number of heterogeneous traffic streams carried by the network grows, there is greater need to study how different types of traffic stream interact with each other as they compete for the same infrastructure.

This chapter considers IP-dial traffic, i.e. customers using their computers to access the Internet by connecting across the PSTN to their Internet service providers (ISPs), using analogue modems or ISDN terminal adapters. The PSTN must attempt to deliver dial tone and connectivity to their PCs, while at the same time protecting the interests of other customers on the network making voice calls.

3.2   Issues

At first glance it could seem that an IP-dial call is, in most respects, similar to any particular voice call. Voice and IP-dial calls are, however, different in a number of important ways. The most obvious is the call hold time. The mean call duration of a voice call is approximately three minutes. The mean call hold time of IP-dial calls depends on the tariff being paid by the customer, but it is significantly longer than that of voice calls. Another key difference is the way in which the calls are initiated. A voice call is dialled by the customer, who listens to the tones delivered by the network to indicate call progress and will converse immediately with the dialled party when the call is connected. When a customer makes an IP-dial call, however, they simply click on the connection icon on their personal computer and the computer monitors the call set-up, possibly redialling automatically in the event of congestion. The customer does not even have to be present; the call will not be lost just because they do not start surfing as soon as the connection is complete.

There are of course other issues, such as traffic volatility (e.g. by introducing a new tariff an ISP may win a greater market share, and therefore the IP-dial load could suddenly shift to another part of the network). While these issues are significant, in this chapter we focus on the fundamental issues of holding time and repeat attempt behaviour. Of course, neither of these are new phenomena — they have been studied many times before and are well understood. Rather, the interest and novelty lie in the mix of traffic and the specific use of trunk reservation in this context, since the magnitudes of the traffic streams involved are so much greater than in network scenarios constructed for those earlier studies.

3.3   Initial Simulation Results

The initial aim of this study was to understand as fully as possible the simplest systems in which IP-dial and voice traffic mix together. We therefore chose to simulate a single link of 600 circuits (ccts) that was offered both IP-dial and voice traffic streams. Simulation is useful as a technique for investigating complex systems because, although it tells us less than a full analytic solution or approximation, the time taken to obtain the results is predictable. Simulation was chosen in this case because the investigation is continuing, looking at more complex systems with multiple routes, overflow routing and traffic controls, which present considerable challenges to any analytic approach.

The actual properties of the traffic streams are complex and rapidly changing in the UK. We assumed mean call hold times of 3 and 30 minutes for the voice and IP-dial traffic, respectively, and that the hold times had a negative exponential distribution. It was also assumed that computers attempting to gain access to the Internet would redial up to ten times in the event of calls failing, and that the redial attempts were separated by 10 sec. This is comparable with the default settings for the Microsoft® Windows® operating system[1].

To simplify the system further, the effect of customer persistence was ignored. This means that if a voice call is lost due to congestion, it is assumed that the user does not initiate any retry attempts and the call is simply lost. In the case of IP-dial, it was assumed that, if the string of ten automatic retries by the PC failed, the customer would then give up trying to connect.

The initial simulations varied the proportion of offered traffic that was IP-dial, from 0.0 to 1.0 in steps of 0.1. The total traffic offered to the system, the load, was also varied, this time from 0.5 Erlang per circuit (Epc) to 1.5 Epc (300 E to 900 E). Each simulation run consisted of an initial 'warm-up' period of 10 hours (simulated time), to ensure the system was in equilibrium, followed by a measurement period of 50 hours (simulated time). Each simulation run was repeated 30 times, with different random number seeds, results being averaged across all 30 runs. The warm-up and measurement periods of the simulations are very long compared with the network-busy period in the actual network. This is because the investigation is concerned with equilibrium behaviour, rather than trying to reproduce busy-hour transients. The warm-up time was chosen to be long enough to allow the system to reach equilibrium. As each simulation run took only a few seconds of computer time, it was not thought necessary to optimise the warm-up.

The standard definition of offered traffic is of course the traffic that would be carried if the system were of infinite size (so there is no blocking and hence no redials). In the presence of redialling traffic streams, however, it is useful also to introduce the quantity which is the total of the mean call hold time × the rate of call attempts. We will refer to these two quantities as the 'underlying offered traffic' and 'effective offered traffic', respectively. Figure 3.1 shows the traffic acceptance ratio obtained for the link at different loads and traffic mixes. We define the acceptance ratio of a traffic stream as the carried traffic divided by the underlying offered traffic. We can see that as the system load increases, the acceptance ratio decreases, but at any particular load the acceptance ratio (and therefore the total carried traffic) is practically independent of the actual mix of IP-dial and voice traffic simulated. This is exactly the behaviour one would expect for the system as a whole.

If we look at the acceptance ratio of the individual traffic streams, a different picture emerges, however. Figure 3.2 shows the acceptance ratio of the IP-dial traffic, and we can see that it is practically 100%, except for when the load is high and the volume of IP-dial traffic exceeds the carrying capacity of the route. The excellent IP-dial acceptance ratio is at the expense of the voice acceptance ratio, shown in Fig 3.3, which collapses in the face of large IP-dial traffic volumes and high loads. It appears that the IP-dial traffic stream always uses the available network resource to full effect while the voice traffic is only able to use the 'leftovers'.

---

[1] Microsoft Windows95 OSR2 defaults: max attempts = 10; inter-attempt wait 10 sec. Microsoft Windows98 SE defaults: max attempts = 10; inter-attempt wait 5 sec. Software diallers provided by ISPs and network operators may, of course, use different values.

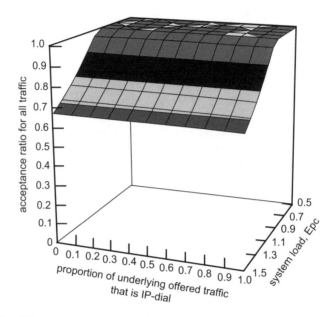

Fig 3.1   The system acceptance ratio as a function of system load and proportion of
offered traffic that is IP-dial.

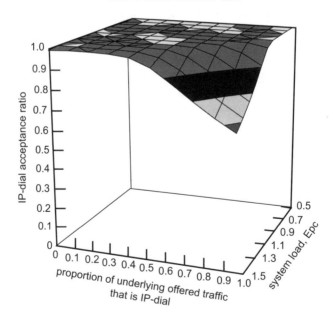

Fig 3.2   The acceptance ratio of IP-dial traffic at various system loads and with varying
proportions of IP-dial traffic offered.

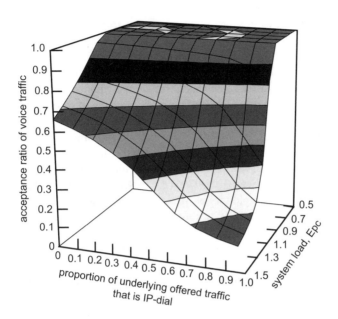

Fig 3.3   The acceptance ratio of voice traffic at various system loads and with varying proportions of IP-dial traffic offered.

When considering these graphs, we should remember that the PSTN is usually operated at loads significantly lower than 1 Epc, and so the large traffic loss we see in the graphs at high loads, is not indicative of normal network behaviour.

This is demonstrated by the fact that a good estimate of the blocking (the probability that a call is lost due to congestion) seen by the voice traffic is that which would be achieved if the voice traffic were offered to a route reduced in size by the underlying offered IP-dial traffic. Obviously, this approximation is only valid if the total IP-dial traffic is less than the route size. In fact, it must be low enough so that it would suffer negligible loss, say 90% of the route size, for reasonably large routes. If we define the underlying offered traffics as T_v and T_d for voice and IP-dial respectively, then for a route of N circuits we can write down our estimate of the voice blocking, B_v:

$$B_v = E(T_v, N - T_d) \quad ; \quad T_d < 0.9N \qquad \qquad ...... (3.1)$$

$$E(A,C) = \frac{A^C}{C!} \left(\sum_{i=0}^{C} \frac{A^i}{i!} \right)^{-1}$$

where $E(A, C)$ is, of course, Erlang's loss formula.

Figure 3.4 compares the voice blocking results from the simulations shown in Fig 3.3 with the voice blocking calculated using equation (3.1). It presents a 1-dimensional view of the accuracy with which the approximation represents the 2-dimensional surface. The points resulting from different proportions of underlying offered IP-dial traffic are plotted using different symbols. For a given IP-dial proportion, we have a set of up to 11 points corresponding to different system loads, between 0.5 Epc and 1.5 Epc in steps of 0.1. (Those points for which underlying offered IP-dial traffic is ≥ 90% of the route size are missing.) We can see that the agreement is quite good, though the few points off the line show that it is not perfect everywhere. The largest 95% confidence interval for the mean blocking measured from the simulation runs was ± 0.007. We can see from Fig 3.4 that the largest deviations between the simple model of equation (3.1) and the simulation results, are significantly larger than the 95% confidence limit and that they all occur at high proportions of IP-dial traffic. This shows that the simple model describes the voice blocking well, except when the proportion of IP-dial traffic is high.

We have looked at the blocking of the voice traffic and have a simple model describing the blocking that the voice traffic sees when it shares the route with IP-dial traffic.

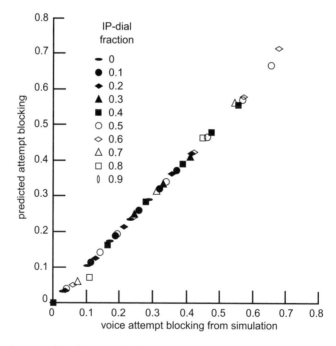

Fig 3.4   A comparison between the simplest voice blocking model and the voice blocking from simulation where the offered IP-dial traffic was less than 90% of the route size.

3.4   Attempt and Intent Blocking

We can see that there are actually two quantities that describe the IP-dial traffic — the attempt blocking and what we will define as the 'intent' blocking. Attempt blocking is the probability that a particular attempt fails, regardless of whether it is the initial call or one of the redials. Intent blocking, on the other hand, is the probability that all of the attempts associated with a particular customer's intention to connect fail. In general, the blocking probability of redial attempts is higher than that of the initial attempt. This is because the redials are correlated with failed attempts, i.e. they only occur if the system is congested. In fact, as the inter-redial attempt time approaches zero, the redial blocking probability approaches unity. Calculating the redial blocking probabilities analytically requires the time dependence of the system to be understood, which is far from trivial.

It should be noted that, because we have ignored user persistence in our analysis, we are identifying each intent with a single attempt to connect to the service by the end user. This is slightly different from the usual definition of intent when considering user repeat attempt behaviour. A result of this is that the intent and attempt blocking for voice calls is identical, because a single user connection attempt (dialling) always generates exactly one attempt. The effect of user persistence is discussed below.

We should also note that there is a simple relationship between intent blocking and the acceptance ratio we defined earlier. The acceptance ratio is simply $1-B_I$, where B_I is the intent blocking.

3.5   Additional Simulation Runs

We have established that the impact of IP-dial traffic on voice traffic is severe when both streams are competing for the same, scarce network resource. The obvious conclusion is that the redialling behaviour of the IP-dial traffic is responsible. To confirm this, Figs 3.5 and 3.6 show the effect on voice blocking and IP-dial blocking, respectively, of changing the maximum number of redials for the IP-dial traffic when the underlying offered traffic is 1.1 Epc (i.e. 660 E). We see that, for a given split between voice and IP-dial traffic, increasing the allowed number of redials increases the voice blocking, until it reaches its maximum, when further redials have a negligible effect.

This threshold shows the number of redials needed to enable the IP-dial acceptance ratio to effectively reach unity. This suggests that it is the redialling behaviour that is the main cause of the collapse on voice acceptance ratio, and that equation (3.1) holds simply because the maximum number of IP-dial redials exceeds the above threshold.

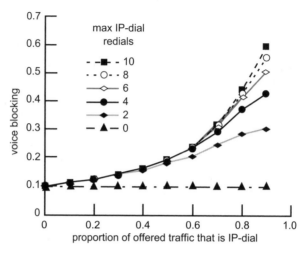

Fig 3.5   Voice blocking versus the maximum number of IP-dial repeat attempts for different proportions of offered traffic that is IP-dial, at a load of 1.1 Epc.

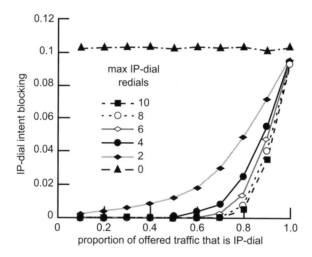

Fig 3.6   IP-dial intent blocking versus the maximum number of IP-dial repeat attempts for different proportions of offered traffic that is IP-dial, at a load of 1.1 Epc.

Does that mean that the comparatively long mean call hold time of the IP-dial traffic has no effect? It is straightforward to prove analytically that mixing calls from two traffic sources, each with Poisson arrivals and negative exponentially distributed call hold times, is the same as having a single source, with total traffic equal to the sum of the two different sources, and the mean call hold time equal to the weighted mean of the two traffic types. Indeed, the independence of blocking from call hold time distribution is a classical result. Figure 3.7 shows the voice blocking with different levels of offered IP-dial traffic. The three curves correspond to IP-dial mean call hold times of 3, 15 and 30 minutes, expressed in terms of the voice call hold time, i.e. 1, 5 and 10 times as great respectively. We can see that voice blocking is independent of the IP-dial mean call hold time except at high proportions of IP-dial traffic, where the voice blocking decreases as the IP-dial call hold time increases. A similar effect can be seen in the plot of IP-dial blocking, shown in Fig 3.8. This apparent divergence from the behaviour predicted by the classical theory is, of course, because the IP-dial attempt arrival process is **not** Poisson.

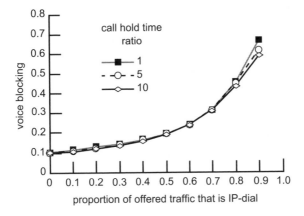

Fig 3.7   The effect of changing the IP-dial call hold time on the
voice blocking at a load of 1.1 Epc.

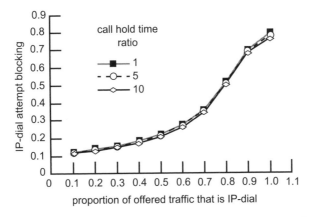

Fig 3.8   The effect of changing the IP-dial call hold time on the
IP-dial intent blocking at a load of 1.1 Epc.

We also considered the effect of the route size, looking at the voice and IP-dial blocking at a nominal load of 1.1 Epc for 300 cct and 1200 cct routes, in addition to the 600 cct route used for the rest of the simulations. These results, after correcting for the efficiency effects associated with different route sizes, showed that there was only a slight effect on the blocking at low IP-dial fractions, but, in common with the call hold time, a larger effect was seen at high proportions of IP-dial — in this case, the larger the route size, the larger the increase in voice traffic blocking at high IP-dial fractions. The call hold time and route size effects are minimal compared with the effect of changing the IP-dial redial behaviour.

3.5.1   Approximate Solutions

As was stated previously, the blocking seen by redials may be different from that seen by the initial IP-dial attempt. The initial calls form a Poissonian arrival process and so the mean attempt blocking probability is identical to the temporal blocking of the route. (The temporal blocking is simply the proportion of time that the route is full.) The redials, however, are correlated with previous blocking events on the route. If the inter-redial time is τ, then the blocking probability for the redial is the probability that the route is fully occupied, given that it was fully occupied at a time τ before. If $\pi_i(t)$ is the probability that the route is carrying i calls at a time t after a call is blocked, then we can write down the following classical simultaneous differential equations that describe the time dependence for a system of N circuits with random traffic (i.e. assuming there are no redials):

$$\frac{d}{dt}\pi_N(t) = \lambda\pi_{N-1} - N\mu\pi_N$$

$$= \mu(\rho\pi_{N-1} - N\pi_N)$$

$$\frac{d}{dt}\pi_i(t) = \lambda\pi_{i-1} + (i+1)\mu\pi_{i+1} - (\lambda + i\mu)\pi_i$$

$$= \mu[\rho\pi_{i-1} + (i+1)\pi_{i+1} - (\rho + i)\pi]$$

$$\frac{d}{dt}\pi_0(t) = \mu\pi_1 - \lambda\pi_0$$

$$= \mu(\pi_1 - \rho\pi_0)$$

where λ, μ and ρ are the call arrival rate, the reciprocal of the call hold time, and the underlying offered traffic, respectively. The boundary conditions are:

$$\pi_N(0) = 1$$

$$\pi_i(0) = 0 \quad ; \quad 0 \le i < N$$

The blocking probability of a redial (assuming that there are negligible numbers of redials) is simply $\pi_N(\tau)$ where τ is the inter-redial delay. Unfortunately, these equations are difficult to solve analytically, but they do suggest the following relationships for a system of fixed circuit group size:

- at a constant offered traffic, the redial blocking will fall faster if μ is large, i.e. shorter call hold times give lower redial blocking at the same load;

- at a constant mean call hold time, the redial blocking will fall more slowly if the offered traffic is higher, i.e. higher loads give higher redial blocking.

If the inter-redial interval is long enough (and μ high enough), we might expect the redial blocking probability to approach the initial call blocking probability. In this state each redial has become effectively de-correlated from the preceding attempt and behaves like an ordinary random arrival, i.e. it will have the same blocking probability as a voice call. In this regime, we can model the system by calculating the effective offered traffic, which is defined as the volume of ordinary random traffic (with no redials) that would result in the same total traffic being carried. This technique is sometimes called the 'modified offered load' approximation.

Consider a system carrying both IP-dial and voice traffic. Let us assume that the inter-redial time is long enough and the hold time short enough for the blocking probability of the redials to be the same as initial call blocking probability, B_y. This means that the redials are effectively de-correlated, and so we can pretend that the redials are simply extra offered traffic. Let us denote the effective offered traffic as T'. We can calculate the temporal blocking, and therefore the voice blocking, B_v, using Erlang's loss formula:

$$B_v = E(T', N) \qquad \qquad ...... (3.2)$$

where N is the route size. We can calculate the carried traffic, T_c, from the blocking:

$$T_c = (1 - B_v)T'$$

But, as we do not actually know what T' will be, by using our definition of the intent blocking, we can write down an alternative expression for carried traffic, this time remembering that the IP-dial traffic will make up to m attempts for each connection intent and that all m attempts must fail in order to lose the intent:

$$T_c = \left(1 - \prod_{i=1}^{m} B_i\right) T_d + (1 - B_v)T_v$$

where T_d and T_v are the underlying IP-dial and voice traffic respectively and B_i is the blocking probability of the ith attempt of a connection intent. Combining the two expressions for T_c, we derive an expression for the effective offered traffic:

$$T' = \frac{1 - \prod_{i=1}^{m} B_i}{1 - B_v} \, T_d + T_v$$

Remember, we are assuming that $B_i = B_v$; $1 \le i \le m$, and therefore:

$$...... (3.3)$$

We can now eliminate T' from equation (3.2) to derive a fixed point expression for the attempt blocking on a route of N circuits:

$$B_v = E\left(T_v + \frac{1 - B_v^m}{1 - B_v} \, T_d, N\right) \qquad ...... (3.4)$$

We are assuming that the $B_i = B_v$; $1 \le i \le m$, and therefore the IP-dial intent blocking is B_v^m.

We now have an expression for the voice and IP-dial attempt blocking, from which we can easily derive the IP-dial intent blocking. Figures 3.9 and 3.10

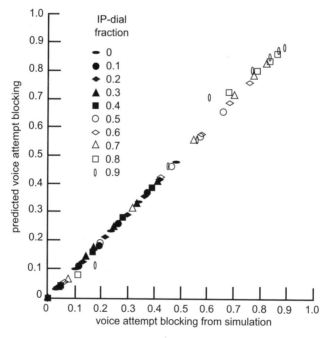

Fig 3.9   A comparison between theory and simulation
for voice blocking.

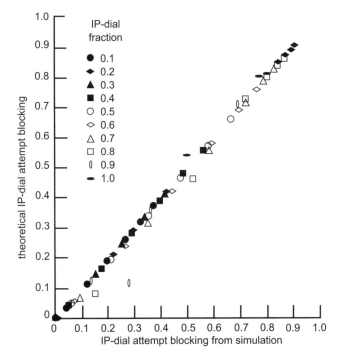

Fig 3.10   A comparison between theory and simulation for IP-dial blocking.

compare the attempt blocking predicted by equation (3.4) with simulation results for voice and IP-dial traffic respectively. In both cases, we assume that the attempt blocking is the same as the temporal blocking of the route.

Figure 3.11 compares the theoretical IP-dial blocking with simulation results. In all three cases the points resulting from different proportions of underlying offered IP-dial traffic are plotted using different symbols. For a given IP-dial proportion, we have a set of up to 11 points corresponding to different system loads, between 0.5 Epc and 1.5 Epc in steps of 0.1.

These graphs (Figs 3.9-3.11) show that the agreement between equation (3.4) and the simulation results is good, except for some points, associated with high proportions of IP-dial traffic, where the difference between theory and simulation is significantly greater than the 95% confidence limits for the simulation measurements. Over most of the parameter space, however, the surface of Fig 3.3 is reproduced well.

In order to see the difference between theory and simulation, Fig 3.12 shows the difference between the two, plotted against the underlying offered load. The data from different traffic mixes can be distinguished on the plot, though data where the underlying offered IP-dial traffic was less than half the total has been eliminated for clarity. The 95% confidence intervals are shown for the 90% IP-dial traffic mix.

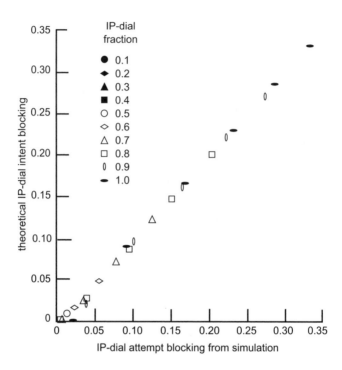

Fig 3.11   A comparison between theory and simulation for IP-dial intent blocking.

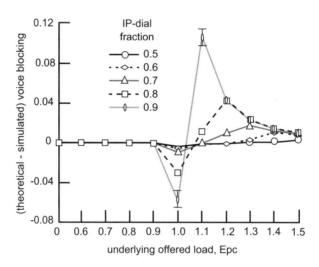

Fig 3.12   The difference between theory and simulation for voice blocking.

We can see from Fig 3.12 that the differences are significant and systematic, but that the magnitude of the differences decreases as the proportion of offered IP-dial traffic decreases. Similar behaviour is seen for the IP-dial attempt and intent blocking.

We have seen that, when the proportion of offered traffic that is IP-dial is 50% or less, the model is a useful tool for estimating voice and IP-dial blocking. To understand why the model performs less well at high IP-dial fractions, we need to consider the key assumption — that the redial blocking probability is the same as the temporal blocking. Although the simulator used did not record the redial blocking probabilities explicitly, it is possible to estimate it using the following expression:

$$B_r = \frac{CB_d - nB_v}{C - n}$$

where, for a particular simulation run, C is the total number of IP-dial attempts, n the total number of IP-dial intents and B_r, B_d and B_v are the IP-dial redial blocking, IP-dial attempt blocking, and voice blocking respectively. Estimates of the redial blocking for the 600 cct route, with a maximum of 10 redials and a 10 sec inter-redial delay are shown in Fig 3.13. We can see that the redial blocking is strongly dependent on the offered load (Epc). At each of the offered loads, there are 9 points corresponding to different proportions of IP-dial traffic, from 0.1 to 0.9. In each case, as the IP-dial proportion increases, the redial attempt blocking also increases.

At high loads, we can see from Fig 3.13 that the redial blocking is very similar to the temporal blocking of the route, as required by our model. So, at high loads, the agreement between the model and simulation is good, as shown in Fig 3.12. As the

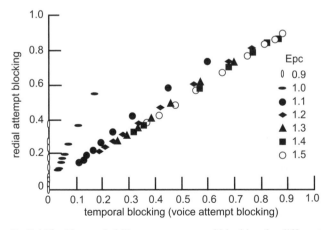

Fig 3.13 Redial blocking probability versus temporal blocking for different underlying offered loads (Epc).

offered load decreases, the redial blocking probability decreases, as predicted by our analysis of the time-dependent blocking probabilities earlier. At high proportions of IP-dial traffic, as the load decreases, the redial blocking falls more slowly than the temporal blocking, however, and therefore the difference between the two becomes greater. This is especially apparent at a load of 1.0 Epc, where, at an IP-dial proportion of 90%, we see redial blocking of 0.55 compared with a temporal blocking of 0.18. This explains the large differences between the model and simulation at 1 Epc in Fig 3.12. At loads of less than 1 Epc, the temporal blocking drops considerably, greatly reducing the number of redials, so even though the redial blocking probability is high, the effect on the system is small, and the model begins to predict the voice attempt blocking well, even for high proportions of IP-dial traffic.

This fortunate feature means that equation (3.4) remains a useful estimate of the expected blocking, even though the underlying model is based on an approximation that holds only over a limited range of conditions.

3.6   So What is a Fair Solution?

We have established that the auto-redialling IP-dial traffic is very efficient at seizing network resource in the event of congestion and that the voice traffic suffers as a consequence. Is this an equitable state of affairs?

There are many possible defensible definitions of fairness, but for the purpose of this discussion, we will define it as follows:

A 'fair' solution is one in which resources are allocated between traffic streams in proportion to the underlying offered traffic.

By definition, if two traffic streams have the same intent blocking, then, during periods of congestion, both streams will lose the same proportion of underlying offered traffic. This means that the network resource will be shared between the two streams in proportion to the underlying offered traffic. It is argued, therefore, that a fair solution is one that matches the intent blocking for different streams.

From a user perspective, this means that a user who dials a number on a telephone once should have the same probability of success as a user who clicks on the Internet icon a single time, even though that click may result in a number of call attempts actually being made.

In order to equalise the intent blocking of voice traffic and IP-dial traffic, it is necessary to protect the voice calls. This can be effected in many ways, some of which are considered below.

What about customer persistence? A customer whose connection intent fails may well try again, and that customer will have a higher probability of getting the telephone conversation or Internet session they desire. It can be argued that this is also fair.

3.6.1 Possible Controls

If we have decided that the allocation of link resource in proportion to the demand is the fairest way of allocating resource in the event of blocking, how could that be achieved? It is necessary to protect the voice traffic from the IP-dial and there are a number of existing controls that could be used in the PSTN to achieve this, and we will consider some of these in turn.

3.6.1.1 Physical Separation

With this technique, interaction between the two traffic streams is prevented by providing separate, dedicated routes for each traffic stream. This guarantees that there will be no unwanted interactions between the different traffics, but does reduce the efficiency of the network. There is a small reduction in efficiency because smaller routes have to operate at lower efficiency in order to achieve the same grade of service as larger routes. There is a more significant efficiency reduction if the IP-dial and voice traffic streams have non-coincident busy periods. For example, on a single shared route, the IP-dial traffic, which peaks in the evening, say, can use circuits required for voice traffic during the morning busy hour. If the traffics are split on to separate routes, that infrastructure reuse cannot occur and additional capacity is required to carry the traffic.

3.6.1.2 Split Route

This technique is similar to the physical separation, except that the separation between the routes is logical rather than physical and it is possible to allow one of the traffic streams to overflow on to the other route. The main advantage of this technique is that the split is not restricted to the modularity of the transmission interfaces.

3.6.1.3 Call Gapping

This technique allows the network operator to restrict the rate at which calls are offered to a particular destination, or range of destinations. It was introduced primarily to protect the switches against call processing overload and there are a number of sophisticated algorithms available to do this; for further details, see Chapters 7 and 8. When using call gapping to control traffic, however, the operator simply defines the maximum rate at which calls may be passed to a particular destination. Calls in excess of this rate are rejected. There are two problems with using call gapping to protect the voice traffic — route efficiency and downstream blocking. The route efficiency problem is similar to that for the separation techniques. In order to protect the voice grade of service, the IP-dial traffic must be restricted to a certain number of circuits (i.e. maximum traffic), leaving enough to meet the requirements of the voice traffic. If the mean call hold time of the IP-dial calls is known, then the maximum traffic may be easily translated into a maximum call arrival rate and the system can behave in a similar manner to the split route. This means, of course, that the volume of IP-dial traffic cannot increase when there is less voice traffic. The impact of changes in the call hold time is even more serious. If the mean call hold time of the IP-dial traffic exceeds that expected when the system was designed, the voice traffic will no longer receive enough protection. If the mean call hold time decreases, then less IP-dial traffic will be carried than expected. The second problem is downstream blocking. In the event of downstream congestion or problems with the destination modem or network access server, calls that fail downstream of the protected link are still counted by the rate limiter as admitted. This may lead to a significant over-restriction on IP-dial traffic.

3.6.1.4   Trunk Reservation

The final control we will consider is trunk, or circuit, reservation. With this control, a reservation parameter is set against a particular traffic stream, which defines the number of circuits that must be free on the route before a call from that stream is accepted. Note that this is not like a split route — the 'reserved' circuits are not real circuits, as the control is merely specifying the highest link occupancy at which calls to a particular destination will be accepted. Trunk reservation is widely used to control traffic interactions in the PSTN — typically, values less than 10 are used. The most important benefit of using trunk reservation is that it only affects call set-up when the route is congested or near congestion. This means that:

- most of the time it would have little effect on the IP-dial quality of service, even if the trunk reservation settings are not precisely optimal for the particular traffic offered to the route;

- trunk reservation allows additional (compared with expected) IP-dial traffic to gain access to the route when there is free capacity nominally provided for voice traffic.

Another advantage is that the trunk reservation is dependent on the end-to-end carried traffic, so that, unlike rate limiter controls, trunk reservation does not have harmful interactions with downstream blocking.

3.6.2   Using Trunk Reservation

Trunk reservation avoids the problems associated with the other possible controls, but will it give operators the control they need? Figures 3.14 and 3.15 display the results from additional simulation runs, showing the acceptance ratio of IP-dial and voice respectively, when a trunk reservation of 4 is applied against the IP-dial traffic. It can be seen that the acceptance ratios of the two streams are much closer and that this system could be regarded as fairer than the uncontrolled system shown in Figs 3.2 and 3.3.

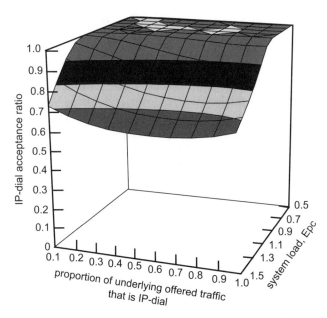

Fig 3.14   IP-dial acceptance ratio for the 600 ccts system when a trunk reservation of 4 is applied against IP-dial traffic.

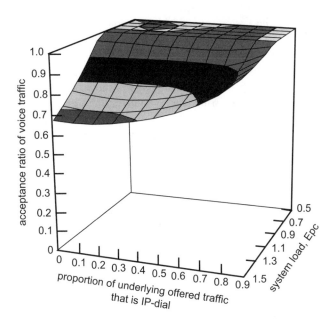

Fig 3.15   Voice acceptance ratio for the 600 ccts system when a trunk reservation of 4 is applied against IP-dial traffic.

In order to understand how a network operator may use trunk reservation to control the interactions between IP-dial and voice traffic at a particular point in the network, we briefly consider network dimensioning and provisioning policies. In principle, knowing the mean traffic matrix for a network makes it possible to calculate appropriate route sizes for all the links in the network so as to meet end-to-end performance metrics at minimum cost. However, for large networks, as such an approach is impracticable, operators may use a set of simple threshold rules that are applied to each type of network equipment.

These thresholds may be expressed in terms of a critical traffic level, which is defined as the highest value that the mean traffic offered to the route can take while still meeting the grade of service criteria set for that route. The calculation of the critical traffic for a particular route will depend on the size of the route, and its location in the network, and should include some contribution to allow for traffic variability.

Other factors, such as the lead time required to increase the capacity of the route, may often be included. In practical terms, however, the critical traffic is often dependent on the blocking experienced at some particular offered load greater than the critical traffic — the overload criterion. Dimensioning to a particular level of overload ensures that the route is large enough to cope with some variation in the underlying offered traffic.

For the purposes of understanding the issues involved in setting controls, we will assume an overload criterion corresponding to 1.1 Epc for a 600 cct route. Obviously, at 1.1 Epc the system is seriously overloaded, and the blocking on the route is much greater than would be experienced at the critical traffic, which is usually significantly less than 1 Epc.

The expected blocking on a 600 cct route offered 660 E of voice traffic is 10.3%. The voice blocking and the IP-dial intent blocking is shown for various proportions of offered IP-dial and different trunk reservation settings in Figs 3.16 and 3.17, respectively.

We can see a fairly complex behaviour, though some obvious trends are apparent:

- at the extremes of the traffic mix, the trunk reservation has little effect on the dominant traffic stream, which, of course, is expected (if there is very little IP-dial traffic, its impact on the voice blocking must be small and so the trunk reservation has little scope for altering the voice blocking) — the same applies when IP-dial traffic dominates, except that the trunk reservation reduces the effective route size slightly so that, even in the absence of voice traffic, the IP-dial intent blocking does increase as the trunk reservation increases;

- in general, increasing the trunk reservation decreases the voice blocking and increases the IP-dial intent blocking;

- the control is relatively coarse as trunk reservation can only take integer values.

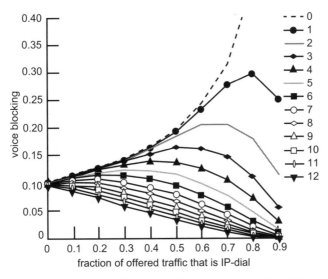

Fig 3.16   Voice blocking versus the proportion of offered IP-dial traffic for various trunk reservations.

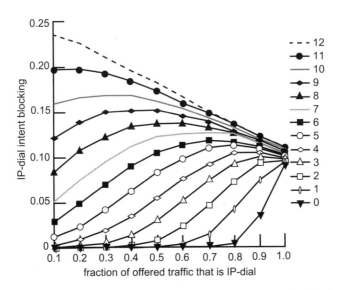

Fig 3.17   IP-dial intent blocking versus the proportion of offered IP-dial traffic for various trunk reservations.

Remember that the objective in applying trunk reservation is to choose settings that equalise the IP-dial intent blocking and voice blocking. Consequently, for a given traffic composition, there is an optimal trunk reservation setting, which can be obtained from Figs 3.16 and 3.17. The optimal trunk reservation as a function of the proportion of IP-dial in the offered traffic is plotted in Fig 3.18, as the bold line (the trunk reservation is on the right hand axis).

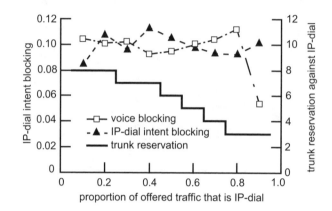

Fig 3.18   A possible arrangement of trunk reservation values for a 600 cct route.

Also shown in Fig 3.18 is the IP-dial and voice intent blocking achieved at each particular traffic mix, with that particular trunk reservation, at a load of 1.1 Epc. Remember that this is the overload grade of service criterion, and so the blocking is greater than would normally be experienced operationally. It is impossible to match exactly the intent blocking of the two streams, due to the discrete nature of the trunk reservation control, but the values shown may be regarded as a possible compromise. Note the interesting behaviour — the **higher** the proportion of IP-dial traffic, the **lower** the trunk reservation required to protect the voice traffic.

3.7   Practical Considerations

We are now able to read off an appropriate trunk reservation for a particular IP-dial voice traffic mix, but we must not forget that the curve is specific to a particular set of traffic properties. We have assumed that the IP-dial and voice streams have mean call hold times of 30 and 3 min respectively and that the IP-dial traffic will automatically redial up to 10 times at 10-sec intervals. But the properties of the IP-dial traffic are not actually known with a great deal of certainty. In fact, the properties are changing as the service matures and new tariffs are introduced. We need to consider how sensitive the system is to changes in these parameters and also the effect of route sizes other than 600 ccts.

We saw earlier that, in the absence of trunk reservation, the intent blocking of the traffic streams depended critically on the maximum number of IP-dial redials. We can see the effect of changing the maximum number of IP-dial redials at a load of

1.1 Epc, when a trunk reservation of 5 is applied against the IP-dial, as shown in Figs 3.19 and 3.20. The voice blocking and IP-dial intent blocking are still heavily dependent on the redial limit. As before, increasing the number of IP-dial redials increases voice blocking and decreases IP-dial intent blocking as expected. The major difference is that when nearly all the traffic is IP-dial and trunk reservation is applied, the voice blocking is low and almost independent of the number of redials. This shows that the trunk reservation is particularly effective in protecting small traffic streams against much larger ones.

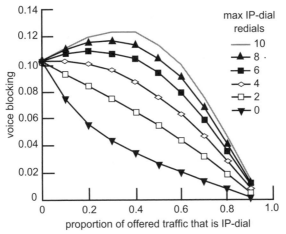

Fig 3.19   The effect of changing maximum number of automatic IP-dial redials on the voice blocking when a trunk reservation of 5 is applied against the IP-dial traffic.

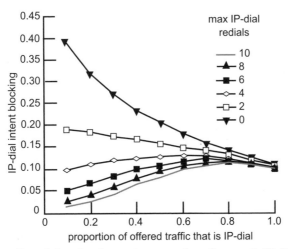

Fig 3.20   The effect of changing the maximum number of automatic IP-dial redials on the IP-dial intent blocking when a trunk reservation of 5 is applied against the IP-dial traffic.

Another factor we considered earlier was the effect of the IP-dial mean call hold time. We saw in Fig 3.8 that the call hold time of the IP-dial made very little difference to the outcome, except, possibly, at very high proportions of IP-dial traffic. Figures 3.21 and 3.22 show how the IP-dial call hold time affects the intent blocking of voice and IP-dial traffic at a load of 1.1 Epc, when a trunk reservation of 5 is applied against the IP-dial, and we can see that the system is now much more sensitive to the effects of changing the IP-dial call hold time.

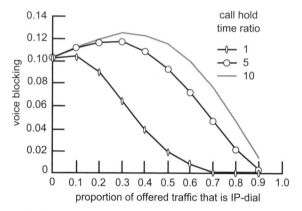

Fig 3.21　The effect of IP-dial call hold time on the voice blocking when a trunk reservation of 5 is applied against the IP-dial traffic.

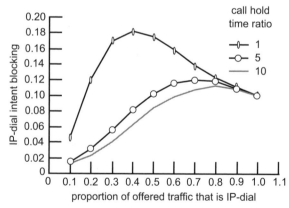

Fig 3.22　The effect of IP-dial call hold time on the IP-dial intent blocking when a trunk reservation of 5 is applied against the IP-dial traffic.

Finally we considered the effect of the route size. We found, that in the absence of trunk reservation, the route size had little effect on the system (after removing the effects due simply to the different system efficiencies). Figures 3.23 and 3.24 show

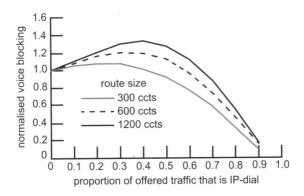

Fig 3.23   The effect of route size on the voice blocking normalised
with respect to an IP-dial fraction of 0.0.

the effect of applying a trunk reservation of 5 to different-sized routes, each being offered a nominal 1.1 Epc. Note that the blocking for the voice traffic is expressed in terms of a multiple of that obtained on a route of that size, at that nominal load, in the absence of IP-dial traffic. This normalisation helps to separate out the behaviour due to the increased efficiency of larger routes, where one would normally see a lower blocking as the route size increases. The IP-dial blocking is also normalised, but, in this instance, to the blocking obtained on the route in the absence of voice traffic.

The graphs in Figs 3.23 and 3.24 show that, as the route size increases, the blocking achieved at a particular trunk reservation is higher for voice and lower for IP-dial. This suggests that on larger routes, a higher trunk reservation is required to protect the voice grade of service — which has been confirmed with additional simulation runs (not shown).

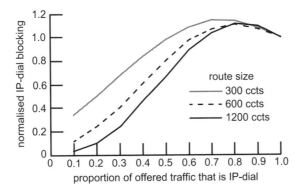

Fig 3.24   The effect of route size on the IP-dial intent blocking normalised
with respect to an IP-dial fraction of 1.0.

3.8 Discussion

We have seen that aggressively redialling IP-dial traffic can seriously affect the grade of service of ordinary voice calls when a shared route becomes congested. (We note that in the absence of congestion there is no unwarranted interaction between the two traffic streams.) As congestion is relatively rare, why does this behaviour require a response from network operators?

Tight control of performance is essential for reconciling network profitability with customer satisfaction, and an important contributor to the latter is the perceived fairness. While emergency calls are already protected against other telephony traffic, through trunk reservation (TR) or other controls, other calls that may still be seen by customers as urgent or essential are not. It is important that these calls are not unduly denied access to network resource during periods of congestion.

This chapter has considered the consequences of adopting, as a criterion of fairness, the equalisation of intent blocking, where an intent represents a single connection attempt by a user, regardless of the number of attempts actually generated by that customer's equipment. Although there are many possible defensible criteria, this has a certain naturalness which means that it is at the very least a strong contender.

Trunk reservation is suggested as the best means to control the interactions between voice and IP-dial traffic, and it has been demonstrated by simulation that TR is able to reduce the difference between the intent blockings of the two traffic streams — and, in the ideal case, to equalise them.

The recommendation of actual values for the TR is complicated by the fact that not only are the properties of the traffic streams highly variable, but their relative proportions can vary greatly from one route, or day, to another. For a proportion of IP-dial traffic in the range 0 to 30%, and with the redial behaviour considered here, relatively high TR may be required, perhaps in the range 8 to 10. This is in addition to any TR required to protect emergency services.

What are the consequences if the selected value is not the optimum? If the trunk reservation is almost right, it will certainly result in a fairer allocation of resources than no trunk reservation at all. This is fortunate, since a network operator can only feasibly design a network to meet average customer behaviour — there is no way at all in which it can be economically adjusted to meet every individual's particular characteristics. In any case, the proportion of time that the route is congested, and the control is active, is low, so that the overall impact of any strategy is limited.

The interactions described in this chapter do not constitute an exhaustive investigation of the interactions between voice and IP-dial traffic, and work is continuing in this area to reconcile the theoretical studies reported here with the very practical considerations of planning and operating a real network.

3.9   Summary

Operators must be aware of the implications of mixing auto-redialling traffic streams with ordinary traffic. The results presented here show how auto-redialling traffic can drastically affect the ordinary voice traffic acceptance ratio under overload conditions. In the absence of congestion, the two traffic streams can be mixed together without undue interaction.

This asymmetry in acceptance ratio can result in an unfair allocation of resources between the two traffic streams.

So, when auto-dialling traffic is competing with ordinary traffic for a finite network resource, it is recommended that trunk reservation is used to ensure, in the event of congestion, a fairer allocation of that network resource between the two streams.

4

TECHNIQUES FOR THE STUDY OF QoS IN IP NETWORKS

S F Carter, N W Macfadyen, G A R Martin and R L Southgate

4.1 Introduction — Approaches to QoS

Following the success of the Internet, over the last five years or so IP has firmly established itself as the networking protocol of choice for a wide range of both traditional and emerging applications. More recently, among business users, there has been much interest in adding real-time applications, including voice over IP (VoIP) and interactive videoconferencing. A major driver for this is the desire to consolidate all applications on to a single, multi-service platform.

The challenge for the network is to meet the very different needs of a wide range of such applications in a cost-effective manner. For example, real-time services have stringent requirements in terms of delay, jitter, and packet-loss, while other applications may be much more tolerant. One approach would be to engineer the network so that all traffic was given the performance required by the most demanding applications. This usually translates into over-dimensioning, which very often is simply not cost effective. Instead, the deployment of appropriate quality of service (QoS) mechanisms within the network allows highest performance to be delivered only to the fraction of traffic that requires this treatment, without the need for overall over-dimensioning. QoS entails the segregation of packets from different groups of applications into a number of service classes, and the provision of different appropriate treatments to these classes at each queueing-point (typically router WAN output interfaces) in the network. The idea is similar to that employed in other network technologies such as ATM, and is not particularly new to IP.

The use of these different service classes is not a question for the network provider, but one for the end-user, and is dependent only upon the applications to be run. The definition of the nature of the QoS of each class is, however, entirely the function of the carrier. Typically, QoS-enabled networks are now converging to some 3-4 classes, of which one is an expedited forwarding (EF) class designed principally for real-time delay-sensitive data; one or two are assured forwarding (AF) classes that give standard data (which is tolerant of delay but not of loss) a greater or lesser priority; and the lowermost class is for standard best-effort data. All except this last class are likely to have service level guarantees (SLGs) associated with them.

The Internet Engineering Task Force (IETF) first began attempts to develop a standardised approach to QoS via the Integrated Service Working Group. Although consistent with the connectionless principles of IP, this approach was based on the notion of multiple service classes combined with end-to-end resource reservation for each individual session (e.g. user-to-user VoIP session) using a signalling protocol called RSVP. Although this IntServ approach found some usage, the approach was considered by many to be too heavyweight and to suffer from serious scaling limitations, and so it has not yet been widely adopted (at least, not in its originally conceived form). The IETF followed this with the differentiated service (DiffServ) approach [1], which began in early 1998. This has quickly become the 'industry standard' approach to QoS in IP networks, and uptake has been particularly great within enterprise networks and provider virtual private networks (VPNs). This chapter focuses specifically on the DiffServ approach to QoS.

DiffServ involves the segregation of traffic into a small number of classes, but unlike either the ATM QoS approach, or the IETF IntServ approach, there is no signalling control plane to look at end-to-end behaviour. Instead the approach relies upon per-class capacity planning at each router individually (the so-called per hop mechanism) in order to ensure that, overall, each class gets the standard of service required.

The three essential components of such an overall control mechanism are given below.

- Classification

    At the 'edge' of the network (e.g. at the output of a router at a particular customer site), traffic is classified on a packet-by-packet basis into one of the designated service classes. Typically an application may be identified by means of static fields within either the IP or the transport layer headers, such as transport control protocol (TCP) or user datagram protocol (UDP). Following classification, the DiffServ code-point (DSCP) field within the IP header provides a means to label each packet according to its class.

- Traffic conditioning (policing and marking)

    It is frequently also necessary to apply some form of policing to some or all of the defined service classes, to ensure that usage remains within prescribed levels. This is particularly necessary for the 'more valued' classes, which receive best treatment, and so have greatest cost to the network provider.

    Policing entails deciding on a packet-by-packet basis whether the packet-stream for a particular class lies within a specified contract profile of mean-rate and degree of burstiness; and if not, taking appropriate action. With VoIP traffic it is common simply to drop packets that are deemed non-conforming; for some data services, however, non-conforming packets may instead at that stage merely be labelled (using DSCP) as 'out-of-contract'. Later in the network, routers will try to forward such packets for a particular class, but under conditions of congestion where some discard is necessary, these packets will be discarded in preference to 'in-contract' packets.

- Differential treatment according to service class

  At each potential queueing point in the network, mechanisms are deployed to ensure that each class gets the appropriate treatment. Typically this is achieved by means of two complementary mechanisms.

The first mechanism entails establishing separate queues for each class and the use of a scheduling algorithm. After each packet has been sent, this algorithm determines from which queue to serve the next packet. A commonly used algorithm is priority queueing combined with class-based weighted fair queueing (PQ-CBWFQ); another is modified deficit round-robin (MDRR). In both of these a priority queue is used for real-time services such as voice, and whenever there are any packets in this queue they are served in preference to anything else. When there are no voice packets to be sent, the algorithm selects the class to serve next on the basis of a configured 'weight' in order to distribute effort equitably.

The second mechanism provides a means of intelligent discard within a particular queue. One application for this is to provide the discrimination between in-contract and out-of-contract packets within a class. The usual means of achieving this is via a queue-management algorithm known as weighted random early detection (WRED).

The main role of the IETF has been to define the DiffServ architecture and develop a common understanding and nomenclature for these elemental building blocks. Much choice still remains with the vendors in terms of precisely how the mechanisms work, and with the network designers as to how they should be configured, and there is much more to designing and building a successful QoS-enabled IP network than simply buying routers which support the necessary features. A natural consequence of QoS is an increased emphasis on the importance of reliable and consistent performance, and operators of such networks are expected to provide service level assurances (SLAs) to describe such objectives. Performance engineering therefore has a key role to play in the successful design and operation of QoS-enabled networks

Application of the control mechanisms outlined above — in particular, that of differential treatment of packets at a router — is, however, fraught with uncertainty. There are many adjustable parameters, and mis-specification of these can destroy the efficacy of a control, or even reverse its effect. It is therefore essential to have an adequate understanding not only of the behaviour of each one in isolation, but also of how they interact in practice.

There are three main techniques available to the performance engineer for this — analysis, simulation, and actual experiment. In terms of speed of application, these range from very rapid (analysis) to painstaking and slow (experiment). Analysis is generally very specific and limited in application, but provides real insight into the fundamentals of behaviour of the effect studied; while simulation is used to study the mutual interactions of effects which are too complex for analytical treatment (of which an obvious example is the behaviour of TCP itself), but generally gives less real understanding of the mechanisms concerned.

Finally, actual measurement and experiment is totally definitive and convincing, and is guaranteed to include all those effects ignored elsewhere for tractability; but suffers from numerous difficulties of both practice and principle. Real experiment is complex, costly, and time-consuming; and is difficult to do adequately since it requires testers which are not only faster and more sophisticated than the systems they are testing, but also have the right statistical properties (an area frequently overlooked). In any case, while testing a single device in isolation may be practicable (and we present an example in section 4.5), testing an entire network in the detail necessary to know optimum parameter settings is obviously totally out of the question. Large systems are not tuned like pianos, by experimental modification until they are right; their characteristics must be understood in an abstract setting, so that optimal decisions can be made without the need for hands-on measurement every time. For these reasons, experiment, although a crucial tool in the armoury of the performance engineer, only augments the other techniques, and cannot replace them.

This chapter therefore gives a simple example of each of these three techniques, as applied to the study and understanding of IP QoS. It will be noted how the detail and the insight diminish as the approach changes. While none of these techniques is adequate by itself, they are all necessary to understand the operation and preferred settings of the overall control mechanisms.

4.2   Behaviour of IP Control Mechanisms

Apart from queueing priority, the principal mechanism available to operators of QoS-aware IP networks to adjust the service offered to different classes of traffic is through deliberate differential packet loss. In the DiffServ framework of per-hop behaviour, this is most commonly done through the operation of WRED, which introduces a (differential) drop rate for packets, depending upon the length of the queue at a router.

The behaviour of the WRED mechanism for practical QoS assurance or control is highly complex, with a large number of interacting dependencies, any or all of which are themselves multi-dimensional and difficult to study even in isolation of the others. The list of these includes:

- behaviour of WRED itself, 'in isolation';

- estimation process used for the queue-length parameter employed by WRED;

- committed arrival rate (CAR) policing process which marks packets as being in or out of contract;

- router queue service discipline;

- behaviour of higher-layer protocols (TCP) which influence the traffic's response to packet drop;

- influence of the remainder of the network — including, in particular, limited-speed access links;

- configuration and set-up of user equipment (e.g. protocol window-sizes);

- traffic characteristics — in particular, any determinism or lack of Poissonality.

The effects of each one of these can be critical to the performance of the system, and every one has numerous degrees of freedom, uncertainties, and configurable parameters.

In addition to all of these, there is the difficulty of deciding on the criteria to employ when optimising or setting parameters — the preferred values turn out to depend upon the (essentially arbitrary) choice of percentage-points of the distributions. That is, the setting which gives the best performance for short transactions may be very different from that required for long; it may be necessary to decide whether to optimise the mean response time, or whether to look at the 95% point, and so forth. These difficulties are not new, but are so intractable that they tend to be overlooked or suppressed because no single best resolution exists.

There is therefore at present a significant uncertainty as to whether the WRED suite of mechanisms does in fact maintain a worthwhile distinction between the QoS of different traffic classes — either in absolute terms ('class 1 gets a noticeably better overall QoS than class 2') or in relative ('At any instant, class 1 always has better service than class 2'). These two statements are not at all equivalent: other work [2] shows convincingly that guaranteeing a variety of perceptibly different absolute service-levels is likely to be impossible, but at the same time the provision of a relative difference which protects one service against another under conditions of congestion is certainly important commercially, and probably feasible in practice.

At first sight, the WRED mechanism is peculiar — setting in place a mechanism to drop packets unnecessarily is counter-intuitive. Its rationale is inextricably tied up with the behaviour of the TCP protocol and the (assumed) behaviour of the network, the argument being that discarding packets from many sources at the same time, as would occur when the buffer was full, might cause quasi-synchronised backoff and re-transmit from numerous sources and hence send the entire network into oscillation.

Whether this is a real danger, as opposed to a theoretical possibility, clearly depends upon the mix of protocols employed and the network characteristics — including, crucially, the mix of round-trip times (RTTs). There is a gradual accumulation of evidence that it may in fact not be.

Whether, however, the WRED mechanisms are really useful is therefore not yet clear. The routers themselves implement QoS separation through their service mechanisms such as PQ-CBWFQ or weighted deficit round-robin (WDRR), and already have an effective means of shedding load (queue buffer overflow, or tail-drop); and it is then a real question whether the complexities of the WRED implementation, with all its parameters that require accurate tuning, are in practice worthwhile.

4.3   Analysis — Queue Length Estimation in WRED

Among the many adjustable parameters available with WRED is the integration time employed for the queue-length estimation process.

The queue size is generally estimated, not by its instantaneous actual value, but rather by a moving average (MA), which introduces both a smoothing and a delay into the system's reaction to increased load; and it is obviously important to know the effect of this.

A particular question of relevance is how to set the averaging period employed in the ongoing MA calculation, and what are the preferred parameter-settings in different operating regions.

The limits of the process are clear — when the half-life (see section 4.3.3) of the estimate is low, the behaviour reduces to the spot queue-length situation; and when it is very high, then the effect of the actual current queue-length becomes very small, and discard will take place at the same fixed rate whatever the queue length, determined by the long-term mean values. What is not clear, however, is the rate at which the results change between these extremes; nor the effects of this on the overall discard profile. While this is but one contributor to the overall performance seen by any traffic stream, it is important for an operator to understand these effects because, without so doing, the multitude of adjustable parameters can be set only by guesswork.

The model of May et al [3] provides a well-known simple approach to analysing the immediate effects of WRED for the case of a Poisson arrival-rate with Poisson service-times, i.e. it is a local model in that it ignores all other network or protocol-related effects. It does, however, assume that the queue-length used to decide whether to discard a packet is just the instantaneous or spot queue length at the time concerned. We study here the effect of relaxing that restriction.

Standing Assumptions

We assume throughout this section, unless otherwise explicitly stated, that:

- the arrival process to all queues is Poisson (i.e. we ignore any bursts of arrivals due to TCP operation or other causes) — this is known to be a good approximation [4] in the core of the network, although less so upon access links;

- all queues have a fixed number of places for packets, independent of packet size — this is of course an approximation, and ignores the detailed structure in terms of (for instance) 512-byte particles of memory with which it is actually implemented;

- the estimate E_n of the queue length is updated upon every packet arrival, whether or not successful, according to:

$$E_n = (1-\alpha) X_n + \alpha E_{n-1}$$

where X_n is the observed instantaneous queue length at that epoch, and α is a parameter in $(0,1)$ — this does not accord precisely with all manufacturers' implementations, but is a useful and sensible place to start (in this section, we are concerned with the effects of varying the parameter α);

while for the example systems studied:

- there are two categories of packets, 'in' and 'out of' contract;

- discard profiles (the probabilities of drop at a given estimator value) are defined by:

$$D(x) \begin{cases} = 0 & x < m \\ = C(x-m)/(M-m) & m \leq x < M \\ = C^0 & M \leq x \end{cases}$$

where x is the current value of the queue-length estimate, and m and M are constant thresholds — the quantities m, M, C, C^0 are all adjustable within their obvious bounds, and generally differ for 'in' and 'out' packets.

For 'in' packets, we assume $M = N$ where N is the buffer size (the maximum queue length); C^0 is of course unity. For 'out' packets, classic random early detection (RED) sets C^0 at unity again, and C at around 10-20%. This, however, produces a sharp discontinuity in the discard profile, and so the ansatz $C = C^0$, sometimes combined with an increase in M — the so-called gentle RED [5] — has been proposed to provide a smoother characteristic.

In the classic case there is a discontinuity at $x = M$; the differences are shown schematically in Fig 4.1. In this chapter we generally assume the gentle behaviour.

Most of the analysis below (but not, obviously, the examples) is, however, independent of these assumptions[1].

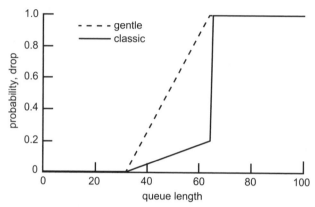

Fig 4.1   RED discard characteristics.

   Although our formalism concentrates for simplicity upon the case of just two categories of packets, which we have assumed to arise through some marking process (using CAR or otherwise), and called 'in' and 'out of' contract; nonetheless it is clear that the techniques used extend immediately to an arbitrary number of categories of packet, which can arise from any combination of DiffServ class of service field and policing marking. The modifications to the analysis below are straightforward and obvious.

4.3.1   Queue Length Estimation

We consider first the distribution of the value of the estimator E of the queue length which is given by an exponentially weighted moving average of the last observations. We assume that the observation process forms a Markov chain, so that we do not have to take into account the previous history, only the most recent value. Let the actual length observed at the observation instant t_n ($n = 0, -1, -2, ...$) be X_n. Then, if the last observation is q, the mean of the estimator is proportional to the expectation of:

$$\left\{ \sum_{n=0}^{\infty} \alpha^n X_{-n} \Big| X_0 = q \right\}$$

---

[1] This is actually slightly different from the original specification, where the initial section remains, and it is only the sharp step which is replaced, by a linear increase ending at the queue length of $2m$. The differences from our modification — subject to a rescaling of M — are negligible.

Now, if the process of queue lengths $\{X_n\}$ is in fact a Markov chain, then the backwards probabilities $B_{j,k}$ are given by:

$$\Pr\{X_{-1} = j | X_0 = q\} \equiv B_{q,j} = P_{q,j} P_j / P_q$$

where P is the single-step forward transition probability matrix; so, the estimator mean is proportional to:

$$= q + \alpha \sum_j j P\{X_{-1} = j | X_0 = q\} + \alpha^2 \sum_j j P\{X_{-2} = j | X_0 = q\} + \ldots$$

$$= q + \alpha \sum_j j P_{q,j} P_j / P_q + \alpha^2 \sum_{j,k} j P_{q,k} P_{k,j} P_j / P_q + \ldots$$

$$= q + \frac{1}{P_q} \left\{ \alpha V_q^{(1)} + \alpha^2 \sum_k P_{q,k} V_k^{(2)} + \alpha^3 \sum_k P_{q,k}^2 V_k^{(3)} + \ldots \right\}$$

where we have written:

$$V_k^{(s)} = \sum_j j P_{k,j} P_j^{(s)}$$

with $p^{(s)}$ the vector of state probabilities at the observation instant $-s$. In the absence of any boundary condition imposed by a knowledge of the probability distribution at some earlier epoch, we equate all these to the equilibrium vector π (i.e. the principal eigenvector of the transition matrix P); and write the expression succinctly in matrix notation as:

$$= q + \frac{\alpha}{P_q} \hat{q} \{1 + \alpha P + \alpha^2 P^2 + \ldots\} V$$

$$= q + \frac{\alpha}{P_q} \hat{q} (1 - \alpha P)^{-1} V$$

where $\hat{q}$ is (the transpose of) a unit vector in the direction of q. The actual estimator required for the queue length is of course $(1-\alpha)$ times this expression, for normalisation.

If service-times as well as the inter-arrival intervals are negative exponential (negexp), two obvious examples where this process is Markovian are those with observations at equispaced time intervals, and those where observations are at arrival instants of demands (whether they are carried or not). The form of the transition probability matrix P naturally varies significantly between these — in the latter case, of course, it is upper Hessenberg (it is zero below its first lower subdiagonal), because the queue length can increase only by 1 between successive arrival instants.

It is difficult to derive the actual distribution of the estimate, which has indeed many interesting properties. On the other hand, a calculation of the variance is straightforward. We observe that:

$$E\left\{\left[\sum_{m=1}^{\infty} \alpha^m X_{-m} + q\right]^2\right\} = q^2 + 2qE\left\{\sum_{m=1}^{\infty} \alpha^m X_{-m}\right\} + E\left\{\sum_{m=1}^{\infty} \alpha^m X_{-m} \sum_{n=1}^{\infty} \alpha^n X_{-n}\right\}$$

$$= q^2 + 2\alpha q/p_q \, \hat{q}(1-\alpha P)^{-1} \cdot V + E\left\{2\sum_{r,s} \alpha^{2r+s} X_r X_{r+s} + \sum_r \alpha^{2r} X_r^2\right\}$$

In the same way as above, the last term becomes:

$$2 \sum_{r,s,i,j} \alpha^{2r+s} B_{q,i}^r \cdot i \cdot B_{i,j}^s \cdot j + \sum_{r,j} \alpha^{2r} B_{q,j}^r \cdot j^2$$

$$= 2\frac{\alpha^3}{p_q}\hat{q} \cdot (1-\alpha^2 P)^{-1} \cdot Q \cdot (1-\alpha P)^{-1} \cdot V + \frac{\alpha^2}{p_q}\hat{q} \cdot (1-\alpha^2 P)^{-1} \cdot W$$

where V is as defined earlier, and:

$$W_k = \sum_j P_{k,j} \cdot j^2 \cdot \pi_j$$

$$Q_{i,j} = j \cdot P_{i,j}$$

This instantly leads to the variance itself.

The most obvious approximate probability distribution to take for the estimate itself is a beta distribution with range $(0, N)$, where N is the maximum value the estimate can take (i.e. the overall buffer size). The beta distribution on $(0,1)$ satisfies:

$$dp(x) = \frac{\Gamma(\alpha+\beta)}{\Gamma(\alpha)\Gamma(\beta)} x^{\alpha-1}(1-x)^{\beta-1} dx$$

with parameters α, β related to the mean m and standard deviation σ by:

$$\alpha = \frac{m^2}{\sigma^2}\left(1 - \frac{m}{N}\right) - \frac{m}{N}$$

$$\beta = \frac{\alpha}{m}(N-m)$$

Obvious modifications apply if the upper bound is not unity.

4.3.1.1   Discard Probabilities

Although the above analysis derived the mean value of the estimate, the usual discard application uses instead a function D of the stochastic estimate X of the queue size, and so, conditioned upon a particular actual state q, its mean value is given by:

$$\text{Pr}(drop|q) = \int D(x)dP(x|q)$$

where $D(x)$ is the discard probability in the state x. If we assume that this takes the form above, then the integral can be expressed simply in terms of a linear combination of cumulative beta distribution functions:

$$\text{Pr}(drop|q) = \frac{CN}{M-m}\frac{\alpha}{(\alpha+\beta)}\{P(s,\alpha+1,\beta)-P(r,\alpha+1,\beta)\}$$

$$+ C^0\{1-P(s,\alpha,\beta)\}-\frac{Cm}{M-m}\{P(s,\alpha,\beta)-P(r,\alpha,\beta)\}$$

where α and β are defined by the mean and variance, and we have introduced the normalised parameters $r=m/N$, $s=M/N$ where N is the upper bound.

4.3.1.2   Transition Probabilities

We start by assuming again that service-times are negexp (this unrealistic assumption will be discussed and relaxed in section 4.3.3), for which the system is totally Markovian. All that is necessary then to complete the analysis is a knowledge of the 1-step transition probabilities of the observation process.

So consider the case of an $M/M/1$ queue, with mean service time unity and arrival-rate a. The probability that exactly k departures occur between one arrival and the next (provided that the queue size at the start is greater than k) is simply:

$$\text{Pr}(k) = a/(1+a)k+1$$

It follows that in this case the 1-step transition probabilities $P_{j,k}$ are just:

$$P_{j+1,k} = a/(1+a)^{k-j+1} \qquad (k \geq j > 0)$$

$$0 \qquad (j > k+1)$$

$$P_{0,k} = 1/(1+a)^{k+1}$$

since the event that 'no departures take place' corresponds to the queue size having increased by one — the previous arrival has augmented the queue.

There remains one correction — an arrival attempt may be rejected, through the WRED mechanism, rather than join the queue. Denote by R_j the reject probability $\text{Pr}(drop|j)$ calculated above for an arrival attempt when the queue length is j; then the effective arrival rate to the queue becomes state-dependent, with:

$$a_j = a \cdot (1-R_j)$$

We assume that the points at which the moving average is evaluated continue to include all those at which an arrival attempt occurs. This change naturally alters the probabilities, which now become:

$$\hat{P}_{j,\,k} = (1 - R_k)P_{j,\,k} + R_k\,P_{j,\,k-1}$$

to take into account the fact that the attempt which started the interval may have been rejected.

We note the alternative and equally plausible scenario, that only *successful* arrivals contribute to the moving average estimate — for a given history, it would result in lower values of the estimator.

Knowledge of the discard rate in each queue state leads directly to an effective arrival rate for each of these, from which the queueing system equilibrium probabilities are easily derived by a simple fixed-point methodology, thus solving the problem.

4.3.2   Examples

The theory above defines an algorithm for the study of the effects of varying the moving-average parameter we have called α, representing the integration period over which the queue-length is measured. As observed earlier, the effects of this can in principle be significant; and in this section we therefore give some illustrations of this, which show that it is indeed so. While the model studied in detail here obviously does not include all the multitude of complicating effects listed in section 4.2, nonetheless it represents another step in the understanding of how these interact, and allows the informed setting of the parameters as opposed to a blind trial-and-error methodology.

4.3.2.1   Standard Parameter Settings

The parameter-space is large, and it is difficult to provide a concise overview. We shall consider as the base point a single queue with 2 classes of traffic which we shall call 'in' and 'out'; and unless otherwise stated, parameters with the settings:

Proportions of packets:	in:	60%	out:	40%
RED thresholds:	in:	(32,64)	out:	(16,32)
Maximum drop targets:	in:	0.2	out:	1.0
Buffer length:	64			
MA parameter α:	0.95			

Figure 4.2 shows the effect upon the discard profiles — i.e. upon the probabilities of dropping an out-of-contract packet through RED, conditioned upon the actual queue-length taking that value — of varying the moving-average parameter α within the range [0.95, 0.999]. The overall offered load is 115%. Gentle RED has been used — the dotted line represents its notional target profile for the out packets, which is indeed obtained when $\alpha = 0$. The curves show a steady progression away from this notional target curve as α increases.

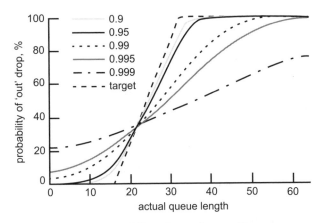

Fig 4.2   Effect of α on drop profiles.

Figure 4.3 shows how the 'out' drop profiles behave as functions of the offered load. In this case, $\alpha = 0.99$, and we have plotted these profiles for load values from 70% to 140%. Gentle RED has again been assumed.

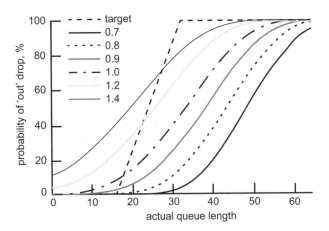

Fig 4.3   Drop profiles and offered load.

The dashed line is the 'target' characteristic, i.e. for $\alpha = 0$. For low loads, the RED discard rates are significantly lower than those for spot observation (no moving average) — this will result in a concomitant increase in the amount of tail-drop over the amount expected, although in absolute terms this may be small. At overloads, on the other hand, an increasing amount of traffic starts to be dropped before the apparent threshold is reached.

4.3.3   Other Packet-Length Distributions

The discussion above has assumed throughout that the service-time (i.e. packet-length) distribution is negative exponential. There were two reasons for this — firstly, it is the standard default distribution, and easy to work with mathematically; but more significantly, it is a necessary assumption for the queueing process analysed to be Markovian.

It is of course not valid in packet networks (see, for example, Chapter 2): fortunately, the analysis above can be extended to other cases, if we assume that the updates made to the moving-average estimate are made at the end of each service; for these form regeneration points whatever the distribution of the packet-lengths (and hence of the service-times).

This extension is non-trivial; it requires a new calculation of the 1-step transition probabilities between one update instant and another, the distinction between the event and time probabilities, and allowance for the fact that tail-drop of packets may occur even when at the last departure the buffer was not full, because it has since then filled. There are also the obvious modifications due directly to the new service-times, and to the fact that now the system state-space is only $[0, N-1]$. Finally, the equilibrium state probabilities of the queue now have to be obtained not through a simple recursive calculation based upon birth-and-death flows, but rather through the extraction of the dominant eigenvector of the transition matrix, which is now lower Hessenberg.

None of these issues presents an insurmountable difficulty. Figure 4.4 shows the queue-length distributions obtained for a buffer of size 32 and 'gentle' RED, and the same remaining parameters as before, for three different scenarios — the previous model with estimator update upon arrivals (labelled negexp (A)); and the new model (with update upon departures) with both 'negexp' and 'Fixed' service-times. Note incidentally the dramatic change that the controls have made to the classical monotone queue-length distribution for a finite queue, which invariably either drops uniformly (for offered traffic which is less than unity) or rises (when it is greater).

Figure 4.4 is drawn for an offered load of 115%. The two negexp models are remarkably similar despite the difference in their estimation processes, which introduces a bias that is at its greatest for a load rather greater than unity (for lower or higher loads than this the models essentially coincide); but the distribution with fixed packet-lengths is significantly different.

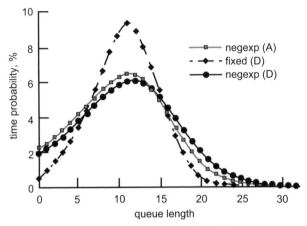

Fig 4.4   Comparison of models.

This suggests that that difference is due predominantly to this change of packet-length distribution, and that (as far at least as this particular view is concerned) we are likely to incur only minor inaccuracy if we replace the one estimate-update process (which is tied to negexp packets) by the other (which is not).

Figure 4.5 shows the overall drop percentage (which is, in practice, concentrated upon the 'out' packets) as a function of α. Total offered load is again 115%, and the minimum drop is therefore 15/115 or 13% of the offered traffic; and the two curves show the total drop for negexp and M/D/1 models (the latter including the tail-drop). We plot this as a function of the half-life H of the observations, that being defined as the number of observations between points which have half the weight: $H = -ln(2)/ln(\alpha)$. The curves extend to the value $\alpha = 0$, which corresponds to instantaneous queue observation and is essentially May et al's model [6]; the dotted lines show the

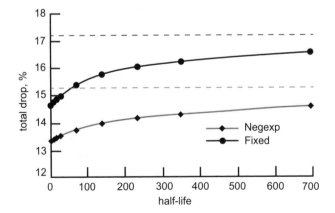

Fig 4.5   Total drop (load = 115%) as a function of half-life.

asymptotes for $\alpha = 1$, or infinite half-life, at which point the observation process is entirely irrelevant and the system bases itself solely on the overall long-term mean and so produces equal discard probabilities in every state.

4.3.4   Discussion

The effects of using a moving-average estimation of queue-length — as opposed to a spot observation — for the purpose of setting WRED discard rates can obviously be severe. The discard rate experienced, which is mostly concentrated upon the lower QoS classes of traffic, can be very much lower than expected if the effects of the estimation methodology are ignored. A knowledge of its behaviour is therefore very important in optimising the settings in practice — although the extent to which that carries through to the overall performance obviously depends critically on the interactions with higher-level protocols and the response of the rest of the network.

Since these settings of the router service discipline are the principal mechanisms available to control the allocation of buffer space between classes, and the way in which any load is shed, they have a crucial part to play in the tuning of QoS-enabled networks. The few examples shown above are sufficient to show that the moving-average parameter plays a significant role, and getting it wrong can largely vitiate the effectiveness of the WRED mechanisms.

Passing to a deterministic service-time, as opposed to a negexp, yields a consistently lower discard rate. The methodology developed is readily extendable to an arbitrary packet-length distribution — the two cases shown here, of fixed and negexp times, have been chosen because they are in a sense the opposite extremes.

4.4   Simulation — End-to-End Performance of TCP

Comprehensive study of the mutual interaction of all the effects possible which contribute to performance can only be done through detailed simulation. In this section therefore we illustrate some of the studies in this area which have been performed through detailed Opnet^{TM} modelling of significant networks. The conclusions of the work, which are not always in agreement with those of widely-published studies which look only at subsets of the contributing factors listed in section 4.2 above, are in good agreement with recently published work of comparable scope from elsewhere [7].

The results presented in this section are based on simulation models for configuring RED at routers [8]. RED was initially proposed to reduce problems associated with the interaction of TCP with tail drop queues; we discuss this here, instead of the WRED of the last section, because the richness of the latter, and the proliferation of parameters and traffic streams involved, makes a concise description impracticable. The ability of such simulation models to evaluate performance from an end-user perspective, which is fundamental to being able to meet customers' expectations, is frequently not available through purely analytic approaches.

4.4.1   Model Description

These simulations were carried out using the Opnet simulation modelling package. Unlike most other reported work, we do not model a fixed number of sessions transferring infinite sized files; instead we model FTP sessions arriving as a Poisson process, with each FTP session performing the transfer of a single file. The detailed modelling is of a single bottle-neck RED queue, which carries traffic from a number of separate sources. These sources generate traffic as FTP file transfers, the file size distribution reflecting that obtained from measurements of Internet traffic in a live network [9]; and in order to reflect the variability found to be present in practice, the round-trip times of the sources covered a range of values from 120-300 ms. While HTTP has not been explicitly modelled, the FTP file-size distribution has been chosen to reflect flow-size measurement data which is typical of its presence [7]. The distribution ranges from frequent small sessions to occasional long ones, and reflects the long-tailed nature of Internet traffic.

The overall effects of RED are extremely complex, and it is not yet even clear whether it is actually beneficial (see, for example, Bonald et al [6]). Different sets of assumptions and models seem to lead to different results; most simulation models in the literature are, however, comparatively small scale, with a few active sessions sending FTP or UDP streams for the whole length of the simulation — the work of Bonald et al [6] considers the highest number of sessions we have encountered, at over 100. We have therefore considered it of the highest importance to construct a model which includes as full and rich a description of the traffic sources as is reasonably practicable.

The basic architecture of our model is shown in Fig 4.6. All sources used TCP Reno with Nagle's algorithm enabled but the SACK option disabled (see, for example, Stevens [10] for a description of this), which employs the fast retransmit and recovery mechanisms in the presence of packet loss. Acknowledgement packets (ACKs) returning from the server do not pass through the RED queue and are not dropped. Packets are dropped from the RED queue based on packet count rather than byte count — the particle structure of real buffers has, as in section 4.3, been ignored. Unless otherwise stated, the other TCP parameters are set to defaults corresponding to Windows 98 for the workstation subnetwork and Solaris 2.6 for the server.

At the top level the model consists of a source of traffic arising from FTP PUT commands, passing files via a RED queue to a server (node_2 in the uppermost box). RED packet drops only occur for packets travelling towards the server — returned acknowledgement packets are therefore not dropped by the RED queue.

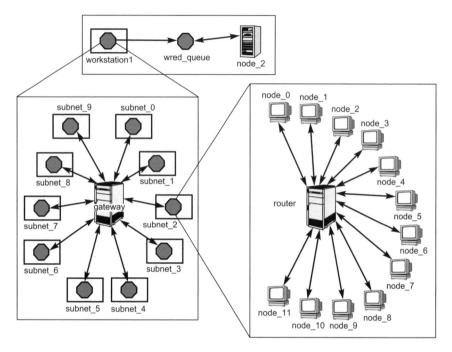

Fig 4.6   Model architecture.

All links in the model are E3 (34 Mbit/s) except for the workstations-to-RED link which has been chosen to be an OC48 (2.4 Gbit/s) so that the RED queue is the only bottle-neck in the system. The one-way propagation delays are 0.1 ms on the link between workstations and the RED queue, and 50 ms on that between the RED queue and the traffic sink. Each of the 10 workstations concentrates 12 separate sources of FTP traffic with a range of one-way access delays between 10 and 100 ms. Each source generates FTP sessions according to a Poisson arrival process, each session transferring a single file whose size is drawn from a distribution derived from the data in Clark and Fang [7].

The server parameters have been specified so that there are no significant delays (apart from the delayed ACK timer) at this node. TCP sessions established with the server are independent of each other (e.g. separate receive buffers are kept for each connection).

The time constant of the moving-average queue length estimator, which, as shown in section 4.3, involves a critical choice of parameters, was chosen to be compatible with the round-trip-time (RTT). In practice, there are a range of RTTs involved, and so an average RTT over all connections must be assumed. The precise details of the estimation process were somewhat different from those of section 4.3, in order to represent more faithfully the actual mechanism employed by the specific routers concerned, but the general principle is very similar. In terms of discard, the maximum value of this (the quantity C of section 4.3, sometimes termed *max*p) was 10% — so that classic rather than gentle RED was used.

4.4.2   Sample Results

The measure of quality of service used in the simulations is the file transfer rate, i.e. the mean rate each user experiences for the transfer of each file. This measure is more useful than the RED throughput, which is much less sensitive to RED configuration and reflects mainly the amalgamated traffic of all users rather than that which each user experiences individually. It is affected by the dynamics of the TCP windowing mechanisms occurring during each file transfer, which in turn is dependent on the packet drop profile at the RED queue — the simulations show how configuring this can have a significant effect on the end-user performance. The results presented here are a few typical examples.

The two main parameters for configuring a RED interface are the minimum (*min*th) and maximum (*max*th) discard thresholds, which in the previous section we called respectively m and M. In this set of results we choose *max*th = 2**min*th as has previously been suggested in the literature. The graphs below (Figs 4.7 and 4.8) show the effect of *min*th on the end user file transfer rate (FTR), which we define as the file size divided by the time to download it. Figure 4.7 considers short TCP flows, consisting of 4 packets, with a bottle-neck RED queue of 256 kbit/s with a utilisation of 85%. The topmost curve shows the mean file transfer rate; below this come the median, the 75%, 90% and 95%. Only the last pair have curves drawn. The dashed curve is the packet drop probability (right-hand scale).

The general trend here is that reducing *min*th has the effect of increasing the drop rate and reducing the queueing delay. Two competing effects are involved:

- packets are discarded earlier — this reduces the mean queue size, and hence the round-trip time, and so the file transfer rate increases;

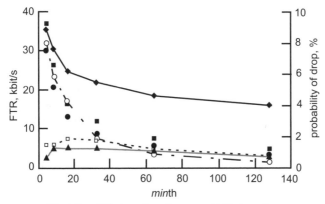

Fig 4.7   File transfer rate — small file transfers.

- packet drop increases — for such small file transfer times, the fast retransmit and recovery process within TCP is unable to operate, since the file sizes are not large enough to generate the triple duplicate ACKs required; packet drop in these cases will therefore cause a retransmission time-out, with a consequential increase in the time to perform the file transfer.

The simulations show that, for short file transfers, and provided that the RED queueing delay forms a significant portion of the round trip time, the reduction in queueing delay resulting from reducing *min*th more than offsets the increase due to retransmissions resulting from packet drops, thus speeding up the FTR.

At the smallest value of *min*th, however, there is an indication that the tails of the FTR distribution (e.g. the 90th and 95th percentile) can be significantly extended. This occurs when the drop rate exceeds about 1.5%. At this drop rate, with appropriate independence assumptions on the packet drop probability, 94% of 4-packet sessions experience no drops.

The appearance of the extended 95th percentile tail here is therefore due to the 5-6% chance that a session will experience at least one drop (and so at least one time-out).

Figure 4.8 shows the effect of *min*th upon large file transfers.

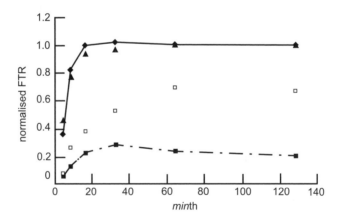

Fig 4.8   Normalised FTR for 16 kB receive buffer.

The file sizes considered here range from 100 to 1000 packets, which in the absence of packet congestion allows the TCP flow-control windows to open fully, to make maximum use of the available bandwidth. The results in Fig 4.8 assume a TCP receive buffer of 16 kB (the default value for Windows 2000). Two RED bottle-neck speeds are considered here, 256 kbit/s and 16 Mbit/s (both at a utilisation of 85%), and the file transfer rates are normalised with respect to the mean FTR when *min*th is equal to 128. The solid curves are the mean and 90th percentile for a 256 kbit/s rate; the isolated points are for 16 Mbit/s.

Reducing *min*th for large files has the opposite effect to that for the small files shown previously. By reducing *min*th, any reduction in queue size (reducing the mean round-trip time) is offset by the increased drop rate, which reduces the mean TCP window size and in turn limits the number of packets that can be transferred per RTT. At low drop rates, then, the fast retransmission and recovery mechanisms of TCP Reno enable efficient use of bandwidth. However, our results show a precipitous drop in the mean file transfer rate (FTR) below certain values for *min*th. This indicates a threshold below which *min*th should not fall if bulk data transfer is not to be severely affected. Our simulations indicate that this threshold is very weakly dependent on the bottle-neck bandwidth.

Further investigation of the model results reveals that this threshold is the outcome of multiple packet drops occurring within a round-trip time when the drop rate exceeds 1-2%, which the TCP Reno recovery mechanisms do not handle very well. Changing the size of the TCP receive buffer significantly affects the value of *min*th where this threshold occurs. Since the operating systems employed by the end users will use a variety of receive buffer sizes (e.g. Windows 98 uses an 8 kbit/s receive buffer by default), this variability will need to be taken into account when configuring such networks.

4.4.3   Effects of Moving Average Estimation Parameter

The simulations were also used to study the effects of changes in the moving-average smoothing factor which in the last section we called α. The routers studied actually make use of an equivalent formulation described as an exponentially weighted moving average (EWMA) factor, which is related to α by:

$$\alpha = 1 - 2^{-EWMA}$$

EWMA varies by integer steps from (typically) 4 to 10, corresponding to an α range of 93.8% to 99.9%; in terms of the half-life $H(\alpha)$ introduced earlier, we have:

$$EWMA = 1/H(1-\alpha)$$

Figure 4.9 shows the effect on the file transfer rate of changing the EWMA, for 1500-byte packets on a 16 Mbit/s link with *min*th equal to 32, for which the recommended value of EWMA is 8 (i.e. $\alpha=0.996$). The points with error bars are the mean FTRs; the others are (from the top) the 50%, 75%, 90% and 95% points. The points on the dashed line are the packet loss probabilities (right-hand scale).

There is a substantial amount of scatter and simulation variability in these particular figures, but the overall conclusion is that varying the EWMA weighting does indeed have a significant effect. The magnitude of this is perhaps surprising, given the relatively small change in overall discard predicted in, for example, Fig 4.5 (although for admittedly rather different parameter values), and emphasises

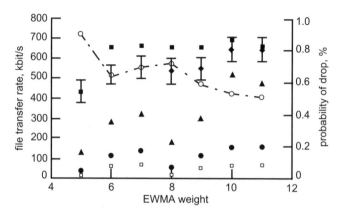

Fig 4.9   Effects of moving-average parameter.

the crucial role played by the behaviour of the higher-level protocol (TCP) as it reacts to packet loss.

Comparison of the figures here with the losses expected using the approach of section 4.3 shows indeed that the packet loss-rate in Fig 4.9 is considerably higher than would be expected. This is basically an effect of the traffic model. Although the mean occupancy here is only 85% (some 29 file transfers in progress), that masks a high degree of fluctuation due to the comings and goings of file transfer requests, and a more appropriate model for the analysis would be a Markov-modulated Poisson process (see, for example, Chapter 2). Because of the sensitivity of the overall packet drop probability to the mean traffic level, the bulk of the discards arise from fluctuations above the mean — which are difficult to quantify analytically, because of the sophisticated adaptive nature of TCP. It is precisely for that reason that recourse to simulation is inevitable.

Simulation results such as these, including the combined effects of WRED and the CAR policing method, have enabled extensive and sophisticated guidelines to be devised for configuring routers in QoS-aware networks. Extensive simulations have identified a wide range of other factors in the system configuration that influence the end-user quality of service provided by these mechanisms, and these in turn have been used to devise practical design rules for multi-service networks.

4.5   Measurement — Validation of QoS Features

The third technique available to the performance engineer is that of laboratory testing. Apart from its validation of existing models, the key advantage of testing lies in its ability to identify effects which have not otherwise been considered, or even known about. Among these are the interactions of the overt performance of protocol and traffic, with the proprietary functional design and performance of a router and its operating system, which are unlikely to be well-known outside the supplier's organisation. It is clear that no other methodology is available for such studies.

Intensive hardware-based testing has always been used before integrating new items into the network; but, with the increasing sophistication of QoS features, a key requirement is to understand their impact and interactions beforehand. Simulation modelling such as that described in the last section is therefore frequently used to guide preliminary hardware testing itself, and in this section we give examples of this. It must not of course be confused with the definitive VV&T (validation, verification and testing) process itself, which is the final stage before implementation.

The laboratory test environment must reflect the live network in which the hardware or feature will be used, but must also allow full stress testing. Commercially available traffic generators that can provide high throughput of various packet sizes are frequently used, but in-house tools are also developed and used to provide more specialised testing of new features not supported by such commercially available tools. In particular, packet analysers are frequently used to confirm particular features that involve changing the bits of the IP packet header. Similarly customer premises equipment (CPE), such as IP and analogue telephones, Web servers and clients, may be used to provide more realistic scenarios.

The range of data to be collected must always reflect the anticipated operational scenario. As an example of this, real-time services, such as voice, create small packet sizes and demand minimal delay, which is very different from a data-transfer application. It is therefore important that a router operating in such an environment should be tested using the buffer sizes recommended for the expected packet sizes and delay requirements. This is illustrated in section 4.5.1.

Testing therefore involves taking many measurements using a variety of tools to indicate the performance of the features and hardware. Results and observations will detail packet loss, queue size, processor utilisation, and results of the packet analyser, e.g. TOS, DSCP classification. Statistics given by the hardware's own self-monitoring processes must always be treated with caution, as, apart from the fact that self-measurement can distort the performance, the actual figures are often provided 'for indication purposes only'.

Multiple tests may need to be carried out in order to understand fully the complex interactions of different features. For example, increasing a system queue depth can lead to higher processor utilisation and reduce loss, but increase the delay. Similarly, larger packet sizes may reduce processor utilisation but lead to queueing buffer fill. A further complication is that apparently uncontentious test settings can, in certain circumstances, produce unstable results unless they are statistically sound, and we describe an example of this.

4.5.1   Channelised E1 Capacity

The first example describes the tests carried out to establish planning guidelines for the number of customers that can be supported when using channelised E1 access to a provider edge (PE) router. The number of ports that can be supported by a router depends on the traffic type (e.g. data, voice) that the customer sends; but the available recommendations on this did not take detailed account of the traffic mix. Direct measurement was not only straightforward, but also the only technique to answer this definitively.

Preliminary testing was performed to define the range of data to be collected during later more extensive tests, and showed that these recommendations needed revision for the particular operating environment envisaged. This revealed too that it was difficult to record an accurate processor utilisation reading, since its measured value fluctuated anomalously during the test. This was identified as being due to an interaction between an insufficiently randomised packet arrival process and the (unknown) internal router sampling schedule; this was confirmed by changing the packet generation mechanisms, and observing that the processor usage statistics then stabilised. It emphasises the importance of ensuring that all tests are carried out using suitably random traffic, to avoid such phenomena, even if the test is not aimed at establishing truly achievable performance figures.

Full tests were carried out using 16 channelised E1 access links — one channel group on each individual E1 serial interface, as shown in Fig 4.10. In order to find the maximum allowable utilisation while keeping the processor occupancy at an

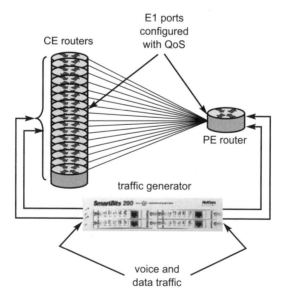

Fig 4.10   Network model for channelised E1 performance testing.

acceptable level, different traffic profiles were used; and systematic studies were also made, varying both the total link traffics and the proportion of the total link bandwidth that was voice.

Table 4.1 illustrates some results from a study of the maximum allowable voice traffic, as a percentage of the gross bandwidth of the E1 links. Data traffic was set in accordance with the standard dimensioning criteria under validation — note that the total traffic is not kept constant from one scenario to another. Voice packets were all 80 bytes and data 512, and in these stress tests the traffic sent in each direction was the same.

The data shown in Table 4.1 represents four sets of measurements, at different nominal voice loadings. We note that for up to 20% of voice, latencies are reasonably stable (indeed, they actually show a decreasing trend) and losses are negligible even though processor utilisation reaches 97%. At the overload point where 40% of nominal bandwidth is voice, however, matters are very different the processor is fully loaded, and both voice and data packets are being heavily discarded. The 'total' figure above shows the combined bidirectional traffic volume, which in light load is just twice the sum of the two preceding cells; the last line, however, shows that under overload the flow becomes asymmetric, due to the operation of CAR on the ingress traffic but not on the egress. The figures shown are the measured egress volumes.

Table 4.1   The effect of traffic mix on router processor utilisation.

Voice (%)		Throughput (pps)			Voice latency (ms)			Processor utilisation (%)
		Voice	Data	Total	min	max	mean	
5	transmit	2265	4123	12 776				
	receive	2265	4123	12 776	0.5	6.1	2.4	68
10	transmit	4341	4017	16 716				
	receive	4341	4017	16 716	0.5	6.4	2.1	85
20	transmit	8859	3587	24 891				
	receive	8859	3580	24 877	0.5	5.4	1.9	97
40	transmit	18 090	2645	41 471				
	receive	9530	2411	19 177	0.5	86.6	70.1	100

These studies confirm the predictions of the theoretical models, and allow the setting of both the maximum allowable proportion of voice and the maximum number of customers on a single PE router, with complete confidence that the system is indeed performing adequately.

4.5.2   Low Latency Queueing Validation

The second example illustrates the validation of a theoretical model. The introduction of a new type of router in an existing QoS-enabled network required a sound knowledge of its characteristics, in particular of the settings necessary for it to deliver the appropriate differential performance. In this case, theory of the system's response already existed, but could not be considered definitive without validation, lest assumptions on the system mechanisms proved to be invalid. Once this is established, the greater ease and simplicity of the theoretical analysis means that no further direct measurement is necessary.

This router employed a modified deficit round robin (MDRR) class-based scheduling scheme to implement low latency queueing. The voice priority queue (PQ) is served whenever it is non-empty, then the MDRR services the other data queues, keeping track of the number of bytes removed from each queue during each scheduling round (the quantum). A large packet held within a queue that was not dequeued in a previous round because its size exceeded the assigned quantum can accumulate quanta during subsequent rounds until the large packet then becomes eligible for dequeueing. The quantum assigned to a given queue on each pass through the cycle is proportional to the (configurable) weight assigned to that queue; and the study focused on validating the proposed mechanism for assigning these.

The model that was tested (Fig 4.11) employed one priority voice queue and three DRR data queues; and the specific queue weights were chosen to ensure good throughput for the data traffic while still meeting the delay targets for voice.

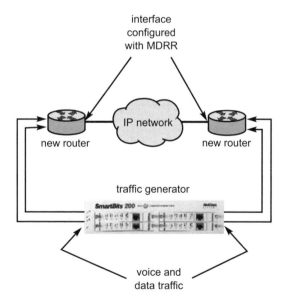

Fig 4.11   Network model for low latency queueing.

In all such experiments, it is important to correct for deterministic latencies within the system as a whole, and the specific item under test in particular. This is easily done using zero-load latency measurements. In the present case, Fig 4.12

shows that a voice packet experienced a delay of about 5 ms in the network even at zero-load, when there are no delays in the priority queue; this minimum delay will therefore also contribute to all the data packets. For the specific case illustrated here, the theoretical model's prediction was that the average total delay of a class 1 data packet, inclusive of this fixed latency, should be 20 ms. Figure 4.12 is in excellent agreement. This case is typical: the theoretical model can therefore be considered validated as being fit for general use.

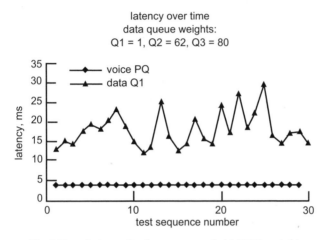

Fig 4.12   Latency for the recommended MDRR weights.

4.6   Summary

This chapter has given examples of the use of the three complementary approaches to network tuning — analysis, simulation, and measurement. The first of these gives in-depth understanding of particular small areas; while the second (with sufficient labour) allows full end-to-end behaviour to be studied — provided, that is, our knowledge of hardware behaviour is sufficiently comprehensive and correct.

The only way, however, to be totally confident of our predictions of hardware behaviour is through actual laboratory measurement exercises. These give definitive and unambiguous results; but are almost always limited in the extent to which they can study full end-to-end behaviour across a network. Moreover, experiments always have to be interpreted in the light of the insight that analysis provides, to be sure that we are in a relevant operating regime and that we have indeed configured the experimental test harnesses sensibly.

An area which is still, however, omitted from such studies is the contribution to network performance that is made by actual user behaviour, i.e. applications usage and characteristics, user repeat-attempt behaviour, the response seen in practice to Web-browsing, etc. Including the detail of realistic traffic demand into a model —

of whatever flavour — is difficult, even when a satisfactory model of traffic exists. Such models can be highly complex (see, for example, Chapter 2) and their application requires experience and skill, whether this is done quasi-analytically, through pragmatic simulation, or through programmable hardware traffic generators. The only machinery available for us to be sure that this is indeed adequately covered is through genuine in-service measurement and testing on actual live networks.

The three methods illustrated in this chapter are therefore all mutually supportive and indispensable to the practising performance engineer. Analytic algorithms for identified components feed into large-scale simulations of networks as a whole; while the results of both of these are confirmed and validated through actual laboratory or field measurement. To close the loop, the conclusions of labour-intensive experiment are then abstracted to a form in which they can be applied readily, in the knowledge that they are both understood and confirmed. Each has its limitations — but together they give a fuller insight than any technique can do separately.

References

1   Heinanen, J., Baker, F., Weiss, W. and Wroclawski, J.: '*Assured forwarding PHB group*', draft-ietf-diffserv-af-01 (October 1998).

2   Gibbens, R. J., Sargood, S. K., Kelly, F. P., Azmoodeh, H., Macfadyen, R. N., and Macfadyen, N. W.: '*An approach to service level agreements for IP networks with differentiated services*', Royal Society Discussion Meeting on Network Modelling in the 21st Century, December 1999 — http://www.statslab.cam.ac.uk/~richard/research/topics/royalsoc1999/

3   May, M., Bolot, J.-C., Jean-Marie, A. and Diot, C.: '*Simple performance models of differentiated services schemes for the Internet*', Proc of Infocom'99 (1999).

4   Cao, J., Cleveland, W. S., Lin, D. and Sun, D. X.: '*Internet traffic tends towards Poisson and independent as the load increases*', in Holmes, C., Denison, D., Hansen, M., Yu, B. and Mallick, B. (Eds): '*Nonlinear Estimation and Classification*', Springer, New York (2002).

5   Floyd, S.: '*Recommendations on using the 'gentle' variant of RED*', (March 2000) — http://www.aciri.org/floyd/red/gentle.html

6   Bonald, T., May, M. and Bolot, J.-C.: '*Analytic evaluation of RED performance*', Proc of Infocom 2000, pp 1415-1424 (2000).

7   Clark, D. D. and Fang, W.: '*Explicit allocation of best effort packet delivery service*', MIT Lab for Computer Science — http://DiffServ.lcs.mit.edu/Papers/exp-alloc-ddc-wf.pdf

8   Christiansen, M., Jeffay, K., Ott, D. and Donelson Smith, F.: '*Tuning RED for Web traffic*', SIGCOMM2000 (2000) — http://www.cs.unc.edu/~jeffay/talks/SIGCOMM-00. pdf

9   Claffy, K. et al: '*The nature of the beast: recent traffic measurements from an Internet backbone*', INET98 (1998) — http://www.caida.org/outreach/papers/Inet98/

10 Stevens, W. R.: '*TCP/IP Illustrated: The Protocols*', Addison Wesley (1994).

5

PERFORMANCE MODELLING — WHAT, WHY, WHEN AND HOW

P Singleton

5.1 Introduction

Performance modelling is the abstraction of a real system into a simplified representation to enable the prediction of performance. It can, however, mean different things to people working in different domains. The basic principles are, in fact, the same, but the people working within each domain have developed these basic principles to best fit their domain's problem space. The two main domains in telecommunications are network performance and IT systems performance. This chapter will focus on performance modelling in the IT systems domain.

Changes in technology, such as third party software, cheap computing hardware, and a movement towards scalable client/server architectures, will neither prevent nor solve performance problems. On the contrary, integration of in-house and off-the-shelf software and the rapid turnaround of development and integration of systems to hit narrow market windows only serve to increase the risk of poor performance. If a system's data and processing become more distributed, then so does its performance risk. The performance of third party software is reduced by its necessarily generic and flexible nature. Rapid developments, using sophisticated development environments that autogenerate code, also create an increased performance risk. All of these risks have to be managed and minimised, making performance engineering, and, more specifically, performance prediction using modelling techniques, an extremely important option for the modern project manager. Performance modelling gives the following benefits:

- relatively inexpensive prediction of future performance;
- design support allowing objective choices to be made;
- decision support for the future of existing systems;

- a clearer understanding of a system's performance characteristics;
- a mechanism for risk management and reduction.

Several performance modelling techniques will be briefly discussed along with some of their advantages and disadvantages. The chapter will then focus on one of these techniques and use a case study to illustrate both how a modelling exercise is carried out, and the benefits it provided to the project concerned.

This chapter will not describe performance engineering and the range of activities it encompasses. What it will do, however, is illustrate where modelling and prediction fits in, and how it interacts, with these other performance engineering activities.

5.2   Where Does Performance Modelling Fit?

The earlier a problem is found, the quicker and cheaper it is to fix. This is a rule that applies to all forms of engineering. Performance engineering is no exception. This does not mean that a great deal of value cannot be added by starting a modelling exercise later in a project's life cycle. It is never too late to use modelling as a decision-support mechanism. Modelling of in-life systems undergoing software or hardware changes, or a changing workload profile, can be extremely valuable. An idea of the type of question that can be answered if started at different stages is illustrated in Fig 5.1.

Figure 5.1 shows the range of performance engineering activities and how they align with an example life cycle (the well-known waterfall development life cycle). Notice that performance estimation and prediction can start as early as system design.

As soon as the first design is formed, the building of a model can begin. Model construction can, however, start at any point within this life cycle and in certain circumstances can only sensibly start when particular components have been produced and tested. If started very early, the model is rough and ready, a first draft, and has a larger margin of error. This is expected. If the model shows that the system capability is an order of magnitude out from its requirements, then this is the time to find out that the proposed solution is simply not feasible.

As the project develops, with more system design information and input data becoming available, then so does the model. Its error margin reduces, the confidence in its results increases, and a more accurate prediction of the system's future performance can be made. The model develops alongside the system.

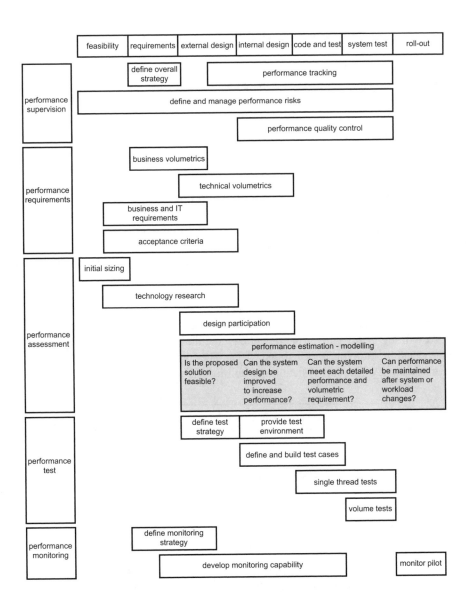

Fig 5.1   Modelling within performance engineering.

5.3   Performance Modelling and Testing

All performance activities can feed information into the modelling exercise. It is worth emphasising, however, the potential for a mutually beneficial relationship between performance testing and performance modelling. To illustrate this we need to look at the level of confidence in the performance of the system gained from different levels of performance testing and modelling. Figure 5.2 shows the increasing level of confidence as the scope of performance testing is widened. It also shows the potential for performance modelling by building on and enhancing the level of confidence gained from performance testing with a reduced scope.

No one is in any doubt that results from a test pushing full transaction volumes through a full-size system would provide the most confidence in the system's ability to perform. Unfortunately, it is very rare that such a system can be made available for testing or that development time is available to write external system stubs and scripts to emulate the full range and volume of transactions that the system has to process. Such a testing exercise can be a massive and expensive undertaking. The lower two testing scenarios, shown in Fig 5.2, are far more common but their results provide less confidence. Modelling can help extrapolate the results from these tests to predict the behaviour of a full-size system under the full future workload. It does this by filling in any gaps in scope not covered during testing. The testing provides the valuable building block input data. The model uses this to construct a representation of the full system under full load. There is no longer a simple trade-

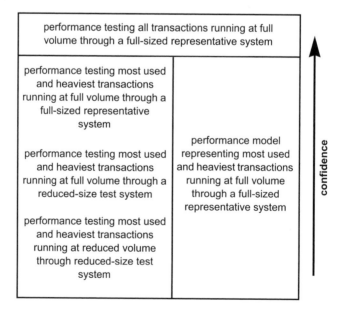

Fig 5.2   Confidence in performance testing and modelling predictions.

off between the cost of a full-size test environment (and development time for scripts and stubs) against the inaccuracy of test results from a smaller system and/or fewer transaction types.

5.4   A Comparison of Techniques

It is worth discussing the main techniques used to model IT systems performance — volumetric analysis, queueing theory and simulation modelling. Figures 5.3-5.5 give a simple pictorial view of the relative strengths and weaknesses of these three modelling techniques. The diagrams are designed such that the more a quadrant is shaded, the stronger the technique is in that area.

Figure 5.3 shows the strengths and weaknesses of volumetric analysis. This is mainly used for capacity planning but can be used for simple throughput capability estimation. One of its main strengths is that it is simple to apply as, generally, only a rudimentary knowledge of statistics is required. In the vast majority of cases a spreadsheet will be used to calculate and present the results. Its main weaknesses are

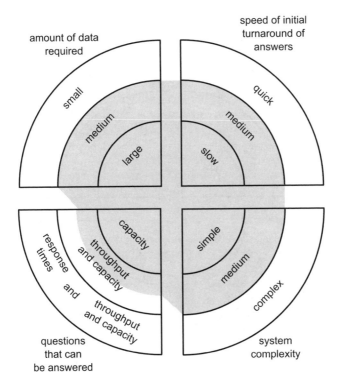

Fig 5.3   Volumetric analysis.

its inability to model complex systems involving conditional routing of transactions, batching, polling and prioritisation. As there is no concept of time spent in queues, this technique cannot be used to estimate transaction response times.

Figure 5.4 shows the strengths and weaknesses of queueing theory. Queueing theory started when the behaviour of simple queued server models was defined mathematically. It has evolved in such a way that a number of queued server types and system complications (e.g. prioritisation, unusual queueing disciplines and combinations of servers) can now also be modelled. It is also possible to string servers together in serial or parallel to form a network and still define key performance parameters and behaviour mathematically, for example using BCMP (Baskett Chandy Muntz Palacios) techniques [1]. This noticeably increased its scope of applicability for modelling IT systems. Unfortunately, it still does not take many system intricacies to break the relatively strict criteria required and make the application of queueing theory extremely difficult. The mathematics involved can become very complex very quickly. However, for very early and simple predictions, especially where a great deal of abstraction is required due to a lack of design information or when the model abstraction of the system meets the criteria required,

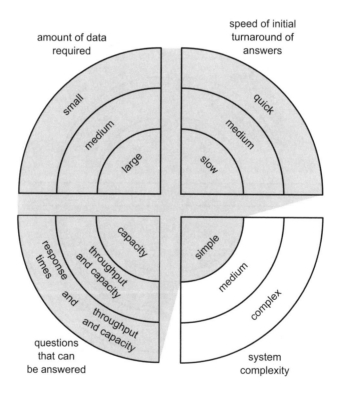

Fig 5.4   Queueing theory modelling.

queueing theory can be invaluable. Once the strict set of assumptions are defined and a model is constructed, results can be produced very quickly using relatively small amounts of input data.

Figure 5.5 shows the strengths and weaknesses of discrete event simulation modelling. Although telcos have been using highly skilled simulation modelling techniques, including the use of bespoke simulation languages, since the 1960s, it is only in the last ten years or so that easy-to-use, GUI-based, software tools have brought the technique out into the mainstream business world. The technique's strength lies in its ability to model any level of system or process complexity. As simulation models have the ability to capture more system complexity, they tend to require more system information and input data to drive them. Obtaining quality input data can be one of the most difficult activities in producing a performance model, and probably more so for simulation modelling.

There are now a large number of software tools available to support each of the techniques mentioned. These tools make it far easier and quicker to produce models. They can also be dangerous for the same reason. When using any modelling technique there is no substitute for knowledge and experience. It is extremely important to understand what a modelling tool is doing for you. Without this

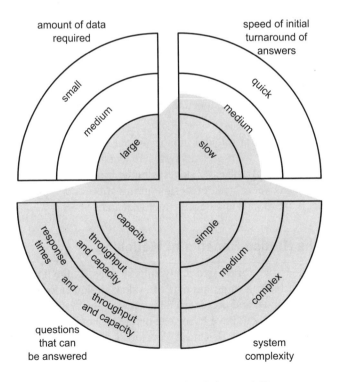

Fig 5.5   Discrete event simulation modelling.

knowledge it is all too easy to accept the results from a tool even if they are incorrect. However, an experienced modeller using a good modelling tool can now produce a high-quality model in a matter of days (depending on system complexity).

5.5   The Modelling Process

The underlying process for each technique is essentially the same. As you might expect, the modelling process has a life cycle of its own:

- model requirements — what questions are to be answered by the model;

- system understanding — gathering all the performance characteristics of the system or process;

- model design and development — translating the system design into a model design, determining what to include and to exclude;

- data collection — usually takes the most (elapsed) time and should be started as soon as possible, e.g. low-level system data not readily available, such as the number of CPU seconds taken by a process to perform a particular task;

- model verification — ensuring the model design reflects the system design by testing the model at various levels and model walkthroughs in conjunction with the system designers, developers and users;

- model validation — if possible running the model to reflect the real system, using the same inputs, then checking the similarity of the outputs;

- 'what if' scenarios, predictions and answers to the questions — exercising the model with specific inputs designed to answer the questions defined in the model requirements.

Usually there is iteration round a number of the stages in the process either as requirements change, or as a greater system understanding is gained, or higher quality data is gathered. At the early stages, simpler models are produced to give an initial prediction. As more system design information and input data becomes available, models are refined and become more accurate.

5.6   The Hidden Benefits of Modelling

The process of modelling can be extremely enlightening. By the time the model has been designed and built, a large amount of system design information will have been gathered, assembled and presented. This view of the system includes an extra dimension that is unlikely to have been seen before — how the system will behave through time. As a system is designed, developed, tested and implemented, this picture becomes clearer. With a modelling exercise running alongside, however, this picture is formed far sooner. A recent example of this was the production of a transaction-level flow diagram for a complete business process. Parts of this information were spread across a number of interface specifications and design documents, but not in the detail required by the modelling exercise. When compiled, this was a view of the process that no-one else in the business had produced and was a view that a large number of people were delighted to see and came to use as an important reference.

A second benefit, gained from using the simulation modelling technique, is the animated performance walkthrough. Most of the current GUI-based simulation modelling tools provide an animation option — extremely useful for verification of the model. However, sit several designers and/or developers in front of an animated simulation and talk through the dynamic behaviour of the system design, and some fervent discussion will almost certainly be generated.

It is sometimes assumed that, if there is sufficient hardware capacity, there are no performance problems. This can prove dangerous as the software or process design can be the bottle-neck. If the system is not meeting its requirements, and CPU, disk I/O, memory, backplane capacity and network capacity are in plentiful supply but are not the bottle-neck, then their plentiful supply is not going to help. In these circumstances throwing hardware at the problem will not fix it. It is worth mentioning that every system will always have a bottle-neck and, by removing one, another is automatically created. The aim is to move the bottle-neck above the requirements, i.e. the system meets its requirements before the bottle-neck starts to restrict throughput and affect response times. A well-executed modelling exercise should help discover where each new bottle-neck will lie and therefore which one(s) would need to be removed for the system to meet its requirements.

5.7   Case Study

The case study focuses on one of the many operational support systems used to sustain BT's business. It will discuss why discrete event simulation was chosen as the modelling technique used, how the design was translated into a model, the questions that the model was designed to answer, and how it answered them.

5.7.1   System Design

The following system description includes those system attributes that are performance affecting. The model does not need to include those parts of the system (or process) that do not affect the performance. Indeed it may also not include parts of the system or process that affect performance only in a small way. The rule of diminishing returns applies. There are a number of factors that affect the accuracy of a model. One of these is the accuracy of the input data, e.g. arrival rate and service-time distributions. The sensitivity of the model results to input data is key to determining the amount of effort to expend on gathering and refining that particular data. The more sensitive the model results are, the more accurate that data needs to be. The skill and experience of the modeller inevitably has the largest influence on the accuracy of model results. Their ability to identify model scope, to translate the real system design into a representative model, and to position effort into the most appropriate model components and input data, will determine the resulting model accuracy.

A brief description of the system modelled in the case study follows. The following acronyms are used:

- QRW — queue reader writer thread;
- RAT process — a line test thread;
- DB — the modelled systems database;
- CSS — customer service system.

The QRW, RAT and the DB are all part of the system being modelled and all reside on the same physical hardware platform. The CSSs are modelled simply as 'black-box' external systems.

The progression of a line test is described below with the corresponding stages also shown in Fig 5.6:

- line test jobs arrive randomly into the system by becoming marked to be line-tested on one of the CSSs (1);
- a QRW thread then reads the line-test job from the CSS (2);
- the QRW thread then writes the job to the modelled system's DB (3);
- a RAT thread picks up the job from the modelled system's DB (4);
- the RAT thread then requests the CSS to carry out a line test [although represented in the model as a simple response-time distribution, the CSS then sends the request to a tester in an exchange, which then attempts to test the line; this test will be either successful (returning a red, amber or green status) or unsuccessful (returning a reason code); there are several reasons for an unsuccessful test, the main reason being that the line is engaged at the time of testing; the tester then returns the result to the CSS] (5);
- the CSS responds to the RAT thread with the line test result (6);
- if the line test is successful, the job is marked as complete on the CSS and is deleted from the modelled system's DB, but if unsuccessful, the job is put back into the DB to be retested in 1 hr (configurable) (7).

The numbers of QRW and RAT threads are both configurable parameters within the model. The components (databases, threads, and interfaces) and the primary process (line testing) define the scope of the model. The request from the CSS to the physical network testers is included as a simple response time distribution, and assumptions, also in the form of a distribution, are made about the success rates of the tests.

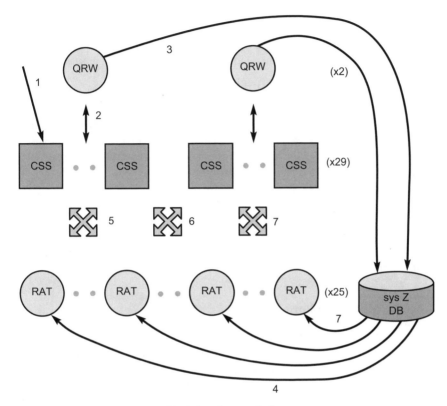

Fig 5.6 System design.

As the system has been live for some time these distributions are well known and relatively stable.

5.7.2 Which Modelling Technique?

There are a number of system characteristics that a modeller should be looking out for to help determine the modelling technique to use. The case study system is an open system and contains a number of system complexities — batching of arrivals (jobs can be bulk marked for testing on the CSS) and polling (the RAT threads can use variable polling intervals). There was also a need to predict the turnround time (response time) for line testing. Volumetric analysis was discounted immediately as response times were required. A BCMP approach [1] was considered. However, due to the system complexities, especially as the RAT thread polling intervals were variable, it was not possible to produce an accurate queueing theory model. Discrete event simulation modelling was therefore chosen.

5.7.3   The Key Questions to be Answered

There were a number of questions and scenarios in which the customer was interested. The key performance requirements that the system had to satisfy were:

- a sustained rate of 850 line tests per hour;
- a mean testing time of less than 2 hours from arrival at the CSS.

These form the basis of this case study. The system was already in place and testing approximately 200 lines per hour. A large increase in workload was expected and the main question was whether the system could cope with this increase while still meeting the mean line-testing response-time requirement. The customer was interested in the following corollary questions.

- Is new hardware required?
- Is a software upgrade required?
- Do some of the system parameters need to be changed to meet the increased throughput?

5.7.4   Model Design

The tool used for the case study was SES/workbench^{TM} [2]. Figures 5.7 and 5.8 are model design screen-shots from the tool and show how the two main pieces of system logic, the QRW and RAT threads, were translated into simulation model logic.

In Fig 5.7, model transactions are created at the node in the bottom left corner (node containing a '+'). As the jobs arrived randomly into the system, the inter-arrival times used a negative exponential distribution with a mean calculated from the required arrival rate. It is important to note that a model transaction does not necessarily represent a system transaction (in this case a line test). Each transaction created in this sub-model represents a QRW thread. Transactions that drop down into this sub-model and appear at the cancellation_css_q and sysz_css_q nodes, towards the top of Fig 5.7, do represent system transactions (cancellation jobs and line test jobs respectively). The QRW thread transactions make calls to other systems, make changes to the database, and use hardware resource. Once each QRW thread is created, it will cycle round the central loop in Fig 5.7 looking for jobs that require testing or cancelling on its set of CSSs. There are 29 CSSs and these are split as evenly as possible across the created QRW threads; for example, if two QRW threads are created, one will scan 15 CSSs and one 14. For each job that requires line testing on a CSS, the QRW thread will loop round the top left loop in Fig 5.7 and enter the job into the modelled system's DB.

Figure 5.8 shows the RAT thread logic. Again, model transactions are created at the node on the left (node containing a '+'). Each of these represents a parent RAT thread. Most of the process logic can be gleaned from following the path the thread transaction takes in Fig 5.8.

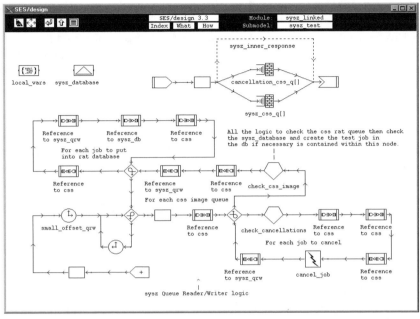

Fig 5.7 QRW thread logic

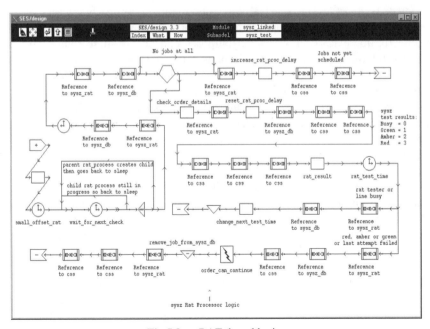

Fig 5.8 RAT thread logic.

Each parent thread loops round the wait_for_next_check node and, if it does not have a living child thread, it creates one. At any point in time a parent RAT thread can have either 0 or 1 child thread in existence. Once created, the child RAT thread then checks the modelled system's DB for line test jobs. If no jobs are found by the child, the sleep delay for the parent thread is increased and the child dies. If the child thread finds a job to line test, it resets the parent's sleep delay to its starting value, and then continues to line test the job.

It was known, from the live system, the percentage of test results that were green (success), amber (not sure), red (failed), or busy (line was not available to test at that time). A number of calls to the CSS are to gather further information prior to testing the line. Although not included in the model, the CSS will, when requested, ask for the line test to be carried out by a physical tester in a local exchange. The throughput capability and response time of the CSS and physical line test system were stable and well understood, and therefore not included in the model scope.

5.7.5   Data Collection

Often one of the most time-consuming parts of a modelling exercise is the data collection. There are a number of different types of data used as input to an IT systems model. Typically these are workload, internal response times, external response times, and model parameters. A workload profile is a set of arrival rates (defined as a set of inter-arrival time distributions) for each transaction type included within the model scope. A model would normally use an existing workload profile (if this exists) for validation purposes and one or more future workload profiles. Internal response time distributions define the times taken to perform basic activities within the model scope and will not include any queueing delays. External response time distributions define the times taken to perform any activity outside the model scope and may represent any amount of complexity and number of queueing delays. Each of these external response time distributions represents a single interface to the model and are sometimes called the 'black box' components within the model. Parameters used in the model are variables, such as numbers of processes, numbers of CPUs, routing probabilities.

The case study system was already implemented, and an abundance of high-quality data was available for use within the model. The response time for the CSS transactions, the response time for the physical line testers, and the percentage of line tests that return green, amber, red or busy, were all available and reliable. The availability of such good quality input data is, however, unusual.

Sensitivity analysis is a technique used to understand how the results from a model are affected by changes to input data. It is standard practice to use sensitivity analysis during all modelling exercises, and it becomes extremely useful when only low-quality input data is available or when particular input data is highly variable. If the model results are sensitive to a particular set of input data, the modeller will make certain that suitable effort is put into ensuring the accuracy of that set of data. An experienced modeller will also be watching for model output that is sensitive to highly variable input data as this brings into question the robustness of the system.

5.7.6   Verification and Validation

Model verification was made using a walkthrough involving the chief system designer and chief system developer. A number of scenarios were animated in turn, using the animation option in SES/workbench, to confirm that the model design was representative of the system design. This included model runs using single transactions sourced at different simulation run times followed by multiple transaction model runs.

As the system is already live, validation was performed by running the model using input data collected from the live system and comparing a number of outputs with equivalent outputs from the model. Several of these validation runs were carried out, each with different workload characteristics. This was achieved by collecting input and output data from the live system at different times of the day and different days of the week.

A number of model runs were then executed using each set of input data and different starting random number generator seeds. Each model output was then compared to the corresponding collected live output data.

5.7.7   Model Runs

A number of scenarios were modelled but only one is included here. The main question the model was required to answer was:

"Can the existing system cope with a sustained hourly rate of 850 line tests while the mean response time remains below 2 hours?"

Figures 5.9-5.11 are screen-shots captured at the end of each model run. They present a set of graphs showing the change in a number of key model outputs as the modelled system moved from time 0 to 100 000 seconds (nearly 28 hours). The graphs illustrate that it took the system several hours to 'warm up'. It is always important to understand the time taken for a system to warm up. The simulation run time will be dependent on the system's warm-up period. Figure 5.9 shows that the warm-up period is approximately a quarter of the simulation run time, or 7 hours. Although this has interesting implications when considering a working day for this system where the line test arrival rate is continuously changing and equilibrium may never be reached, these are not considered here because the objective of this modelled scenario was to determine if the system could sustain a particular hourly rate.

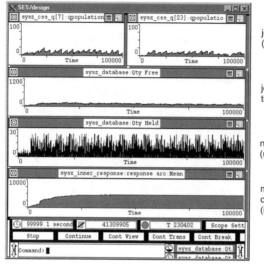

jobs in the CSS waiting to be tested
(2 randomly selected from 29)

jobs in the system DB waiting to be
tested

number of busy RAT threads
(utilisation of the 25 available)

mean time in seconds to
complete a line test and inform the CSS
(in seconds from beginning of model run)

Fig 5.9   Results running at 200 line tests per hour.

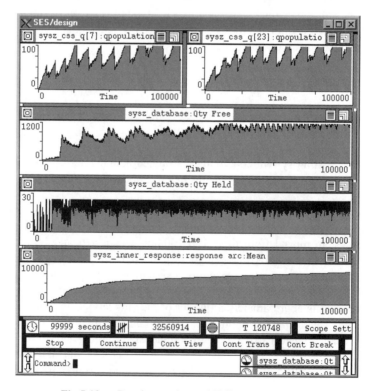

Fig 5.10   Results running at 850 line tests per hour.

The first of the screen-shots, Fig 5.9, also contains a brief description which is applicable to each graph.

Figure 5.9 shows that the system was in equilibrium and coping well with 200 line tests per hour as none of the graphs continue to grow over time. The upper three graphs show queue sizes, the top two showing the number of jobs (queueing) on 2 of the 29 CSSs. The third shows the number of line tests in the modelled system's DB waiting to be tested. The saw-tooth effect, seen in the top two graphs, is due to periodic scanning of each CSS by a QRW thread before marking a number of the jobs as 'under test' on the CSS and entering them as jobs to be tested in the modelled system's DB. The periodicity of the saw-tooth was an interesting effect of the system design and is dependent on the number of QRW threads and the line test job arrival rate. It was highlighted for the first time by the modelling exercise.

Figure 5.10 shows the system running at 850 line tests per hour. It is clear that the system is far busier. The fourth graph in Fig 5.10 shows that the 25 RAT threads are performing tests most of the time and therefore have a high utilisation. The system does settle down and reach equilibrium, however, even though queue sizes on the CSSs and in the modelled system's DB are noticeably higher than the previous

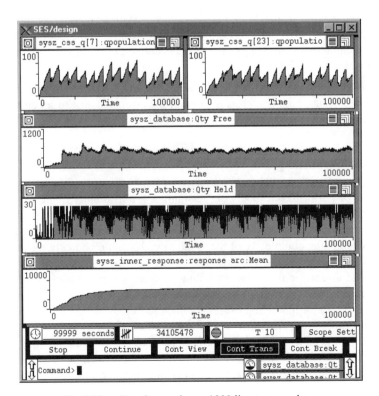

Fig 5.11   Results running at 1000 line tests per hour

model run. The response time still remains stable at a mean of approximately 5000 seconds — less than the 2-hour requirement.

Figure 5.11 shows the system running at 1000 line tests per hour. It is clear that the system is not coping and equilibrium is not achieved. Queues begin to grow, the RAT threads have a very high utilisation (close to 100%) and the mean response time continues to grow. Results from this run clearly show that the system could not cope with this line test arrival rate.

5.7.8   Model results

The modelling exercise provided the customer of this work with the following:

- a clear understanding of their system's dynamic design, i.e. how their system behaved as it moved through time;
- a far better understanding of their system's performance characteristics and to which parameters the system is sensitive;
- assurance that the system would be able to sustain an hourly rate of 850 line tests while meeting the response time requirements;
- assurance that the system would not be able to sustain an hourly rate of 1000 line tests or more — a rate of 850 line tests per hour is close to the limit of the system (if the requirements on the system increase above this, then significant changes will need to be made).

This all adds up to very strong decision support. The cost of the modelling exercise in this case amounted to a fraction of the estimated cost to set up and carry out volume testing to answer the same questions.

5.8   Summary

A number of performance modelling techniques are available. Some of the main techniques that can be used to predict performance of IT systems and networks have been discussed. Each has its own merits. Which to use is dependent on the questions to be answered, the complexity of the system involved, how quickly the results are required, and the availability of relevant modelling expertise. Modelling tools can be very useful in speeding up the modelling process, but care must be taken if used by an inexperienced modeller. There is no substitute for knowledge and experience when carrying out a modelling exercise. Model accuracy depends on a number of criteria but is dependent primarily on the ability of the modeller.

Although it is better to start modelling as early as possible, a modelling exercise can provide valuable decision support throughout the system's development and when the system is live. There are a number of benefits a modelling exercise can provide when it discovers that a system will not meet its performance requirements.

A number of design changes can be modelled to determine their relative impact on the performance of the system. By the use of mathematical techniques or a sequence of simulation runs using incremental parameter changes, it is possible, by iteration, to move towards an optimal design. The modelling exercise can easily move from a role of decision support to problem solving and back again. If started early enough, it is also possible for several different designs to be modelled such that the results can show the trade-off between cost and performance. This enables an objective and effective performance design choice right at the outset.

The case study, although relatively simple, showed the importance of prediction. Without this prediction there would have been no view on the system and process throughput and response time capability under the future increased workload. Should a new system be bought? Even then, would it provide the required capability? Is the software design scalable and to what point? These are questions that can only be answered by modelling, testing, or 'wait-and-see'. In this case, the risks of 'wait-and see' were too high and the cost of testing was far higher than that of a modelling exercise. As the system was already live and running with lower volumes, making key information available, a modelling-only solution to prediction was used. The exercise was quick and provided simple answers to the otherwise difficult questions. Modelling can be a very cost-effective way of providing great benefit.

References

1   Baskett, F., Chandy, K. M., Muntz, R. R., and Palacios, F. G.: '*Open, closed, and mixed networks of queues with different classes of customers*', Journal of the ACM, **22**(2), pp 248-260 (April 1975).

2   '*SES/workbench*'[TM], HyPerformix UK Ltd, Wharf House, Wharf Street, Newbury, Berkshire RG14 5AP — http://www.hyperformix.co.uk

6

PERFORMANCE TESTING — A CASE STUDY OF A COMBINED WEB/TELEPHONY SYSTEM

J C C Shaw, C G Baisden and W M Pryke

6.1   An Introduction to Performance Testing

In its most abstract sense performance testing is the loading of a system, or a part of a system, with a synthetic workload, i.e. taking a real system and using specialist scripts and tools to emulate the activities of multiple users (where a user may be another system) performing typical user actions. Atypical workloads are sometimes used to stretch a system to, and beyond, the anticipated limit to add confidence that it will be able to support, for example, a sudden surge in workload volumes stimulated by marketing activity. This subset of performance testing is known as stress testing.

Performance testing may have several objectives, for example:

- to identify bottle-necks and determine the optimum configuration;
- to determine the capacity of a system end to end;
- to ensure the system functions correctly when under load;
- to investigate what happens when the system is overloaded;
- to evaluate new products or components.

These are the familiar objectives of all performance engineering work.

What differentiates performance testing is that the focus of the work is on a real system with a controlled load rather than a paper or computer simulation-based exercise, or a live system with a live and uncontrolled load. An advantage that performance testing has is that it works with the full complexity of the system in its implemented state, rather than a simplified model, or how it should work according to the design. The downside is that testing can only begin when there is something to test and this is inevitably later in the life cycle than other forms of performance engineering, and obviously major problems found at a late stage are more costly to fix than if they had been found at an earlier point in time.

However, as many modern systems are an integration of specialist components bought in from third parties, who may be reluctant to share details of how their product operates for good commercial reasons, performance testing may be the first opportunity for the integrator to find out if the components being used to build the system live up to their specifications and expectations. It should be noted that performance testing is not limited to a particular technology, although the choice of tools used may be strongly linked to either the underlying technology or to the system interfaces where work is presented.

6.1.1   When to Performance Test

The maxim of all testing — 'test early and often' — applies equally well to performance testing. However, performance requirements can be forgotten, ignored or seem to be of secondary importance during the early stages of product development and sometimes are only considered as a final pre-deployment check. Performance testing becomes that check. In other situations, performance requirements have been considered from the outset and modelling has been done, but there comes a point where people want to see what the performance of the product will be prior to deployment. Ideally this occurs at a component level first, as well as covering the end-to-end system. A further, increasingly common, scenario is that systems are a complex integration of the best components available and it is important for service providers to demonstrate the completeness of the end-to-end solution before launch. However, there are a great many pressures on service providers, and therefore also on system vendors, that can have an impact on the degree of performance testing at various stages, as described below.

The implosion of technology markets seen in recent years has forced companies to drive harder than ever for market share. The result is that there is great competition for valuable customers and, within a highly competitive market-place, the only real way to survive is to offer new, value-add services to attract, and keep, a high volume of customers.

However, high volume platforms are not cheap, they use state-of-the-art technology, often built around third party software and hardware components as soon as they are available. The need for products to be in the market-place quickly, bringing the next generation of service, means deploying new, faster, higher capacity service platforms without delay. Service providers are demanding delivery from system vendors who, in turn, are demanding delivery from hardware and software suppliers, and so on, down the supply chain.

When the next generation of software, chip, card or platform first hits the design stage, the vendor must build the next generation of test facility to put it through its paces. Since this next generation of chip, card, application server or platform is the biggest and fastest around, there may not be suitable components available to build that test facility. Further pressure is exerted to keep costs to a minimum, yet expensive test facilities force the product price up.

Under pressure to get designs on to the production line, many vendors therefore choose or are forced not to perform full performance testing on their product. Commonly they choose to perform 'representative' tests. Examples of these are medium-volume testing with extrapolation of results, simulation of high loads while sample testing, etc. This approach is inevitably non-exhaustive and does lead to products which cannot perform to their design specification. For example, a card vendor must rely on the chips they use to perform at their specification limits, as the platforms rely on the cards.

The further up the system chain we get, the more these performance testing problems are compounded. Service providers are at the end of the chain facing the fee-paying customer. Service providers' reputations rely on the perception by their customers of the whole system and so service providers rely on the systems procured to meet their design specifications — often when it cannot be proved they do. So when is performance testing useful? Whenever your reputation and business survival depends on it!

Performance testing could therefore be the first time in product development that performance engineering skills have been brought into play or could be the final stage in demonstrating the abilities of a thoroughly engineered solution. If performance testing is the product's first exposure to performance engineering then the performance tester's tasks will include those other activities that are usually associated with earlier stages in the performance engineering life cycle, for example:

- identification of the requirements and testing objectives;

- analysis of the application — its components, configuration and key work-flows;

- determination of user and transaction volumes.

If performance testing happens as part of a comprehensive performance management exercise, the testing elements would interface with the requirements, etc, of the previous phases to help in devising the test strategy, writing scripts, building a test environment, building and loading realistic volumes of test data, etc.

While a system is being tested it is important to record as much information about how it responds to the applied workload as possible. While measuring system response time and error rates will indicate if there is a critical bottle-neck, information about system resource usage, database queries, etc, will help identify where the bottle-neck is and so where development effort should be focused.

Bottle-necks can occur in the performance testing equipment as well as the system under test, and, if this occurs, the results recorded are useless as the performance of the system will be hidden by the failure of the test kit to record the response of the system accurately. Therefore, monitoring of the load testing system itself should not be overlooked, especially if complex scripts are being used as these inevitably demand more resources than a simple script.

6.1.2   Tool Choice

Modern systems come equipped with many interfaces, not only plain old telephony service (POTS) network connection interfaces (integrated services digital network (ISDN), CCITT signalling system number 7 (C7), etc), but also World Wide Web (WWW), messaging and computer/computer interfaces. This means several tools are required to fully exercise such systems as they are expected to be in a live environment. This can also mean different test expertise is required, drawn from different teams or areas of speciality.

The decision as to which tools to use should be driven by which work best with a particular system. Certainly with Web applications, although all tools can adequately exercise a simple application, there are marked differences in the capability of tools in their support for the latest Web technologies that developers like to include in their products. This is exacerbated when Java^{TM} applets or Active X^{TM} controls are used that work within a browser environment but do not restrict themselves to the hypertext transfer protocol (HTTP). However, it is also true that the various tools rapidly catch each other up and leap-frog one another in abilities and features.

The ability of a tool to support a given interface is key, but other factors to consider in choosing a tool are the scripting interface, flexibility of control (ramps, mixed workloads, etc) and timing accuracy, as these will help to reduce both the preparation time, and the time required to execute the tests and collate the results. As with a lot of performance engineering tasks, finding the information from the data is a central part of the work, and a good reporting feature set is a great asset. It is also vitally important to understand whether the tools generate an open (asynchronous) or closed (synchronous) workload as this will have an impact on their ability to mimic the arrival patterns of real traffic. Tools that wait for the completion of a work request (e.g. fetching a Web page) before submitting another, generate a closed workload which smoothes and so reduces the instantaneous load offered by tailoring the arrival process of generated messages to the stochastic delays experienced within the system under test. This has the effect of stabilising the system under conditions of congestion and so prevents it from being overdriven as can happen in a live situation — this is a well-known mechanism for producing spurious (optimistic) results if not accounted for. Another, often overlooked, aspect in constructing a valid test is the impact of network delay between the generator and system under test. While there are advantages in placing load generators close, in network terms, to the system under test, this reduces the load as messages are acknowledged speedily, reducing the number of messages the system is handling at a given time. This in turn reduces the memory requirements of the system. BT's longstanding expertise in performance testing allows full assessment of the implications of tool behaviour and environment on test validity, and enables corrections or test modifications to be made where necessary.

Performance test tools are a significant expense for a single project to bear and the sophistication of the best tools, while on the one hand making initial scripting simple, also means that they require significant expertise to configure and operate optimally. These cost and complexity factors combine to make the use of specialist performance testing services attractive as they will:

- have access to a variety of tools, enabling the best to be selected;

- be able to supply highly trained and experienced people to run the tests;

- charge for tools on a usage basis (i.e. reduce cost).

These advantages help reduce the time to performance test, improve the accuracy of the results and relieve projects from the choice of using a second rate tool or bearing a significant capital and maintenance cost.

6.1.3   Pitfalls

What follows are some examples of perhaps less obvious pitfalls.

As described above, cost, practicality and market pressures combine to demand rapid deployment which results in platform specifications being virtually impossible to fully test. It is a brave company that stakes its reputation on these specifications, and experience has shown in almost all cases that such specifications must be tested or verified rather than trusted.

As service providers in a very competitive market-place seek to provide new, innovative services, designed to attract new customers, it is unlikely that platform vendors will have fully considered the applications which these service providers intend to deploy on the platform. Service providers must not therefore assume that vendors will have fully tested the system under the conditions to which it would be subjected in live operation, an example being how application-logging activity, telephony call rate or mix of call types may affect the database performance, with a knock-on effect on Internet and/or telephony performance.

As mentioned previously, a modern system often sports a collection of interfaces. These different interfaces will inevitably be linked and have some level of interdependency within the system under test. It is therefore folly to conduct performance testing on less than all interfaces at the same time. It may sometimes be justified to measure performance at a single interface, but all interfaces must be exercised across their full work ranges to ensure interdependency testing.

Black box testing is seldom appropriate when performance testing. Simple problems either internally to the system under test or at the system interfaces can generate warnings, possibly in the form of alarms, that in themselves are not service affecting. A large volume of them, however, can have disastrous effects. The rate at which they are generated or large numbers accumulated over a given period can fill up disks, take up system bandwidth, etc; if not disastrous, these problems can limit overall system performance. It is prudent to monitor system activity at interfaces and internally for expected behaviour during performance testing.

There may also be unexpected behaviour between systems. Systems that operate complex protocols across their interfaces can operate in a number of different, configurable ways. These protocols are sometimes clever and flexible enough to cope with small incompatibilities between these interfaces. However, when the system is about to go live, configurations are defined, by separate teams, as part of the installation process. In this case the live configuration may not have been that used for performance testing which can have a significant impact on the validity of the test results. It is therefore important to analyse configuration details to ensure the integrity of the performance test environment.

Further complexities arise when the test model is not the same size as the live system. Again this is a cost-driven factor as full systems are often very expensive and consequently not purchased and installed until a short time before 'going live'. The best that can often be hoped for is representative operational readiness testing. It may be difficult to justify sufficient time to perform performance tests on a live system prior to launch and the risk of post-launch testing may be too great.

Inevitably this means that sometimes performance tests have to be run on a scaled-down model. In this case care needs to be exercised to ensure the test environment exhibits as near identical architecture to the live model as possible. This approach will necessitate performance results having to be extrapolated at the very least, and is therefore less than ideal as this simple transformation seldom accurately models the effects resulting from the resizing of a complex system.

6.2   Case Study — *evoca*

evoca has been developed by BT. It is the result of work looking into the disruptive business models of the virtual service provider (vSP). Simply put, the vSP re-sells connectivity purchased at wholesale rates from existing providers, adding value and differentiation through integration of applications focused towards their customer base or chosen market segment. The key concept of *evoca* is a set of integrated application components that enables the combination of fixed voice, mobile voice and Internet access authentication and services. Figure 6.1 represents the building blocks of the *evoca* solution.

A key element of the *evoca* business solution is its ability to give its customers the opportunity to focus their marketing resources on the value added applications as shown on the right hand side of Fig 6.1. However, in order to achieve this, the left hand side elements in Fig 6.1 are needed and indeed are expected as part of the base platform. It is the integration of these key elements which is considered in this chapter.

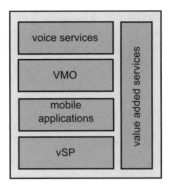

Fig 6.1 *evoca* building block concept.

The starting point of the *evoca* solution is the 'virtual service provider' management environment. This logical area is the heart of *evoca* and provides the Internet service and applications management. This includes the usual virtual ISP facilities — on-line registration, customer self-care and account management, as well as Web hosting (for the platform operator), e-mail accounts, etc.

The platform capability includes gateways for mobile applications. SMS and WAP are used as the basis for delivery of services such as e-mail, news, mobile Internet access and applications.

Within the *evoca* architecture there is the provision for the platform to act as a virtual mobile operator (VMO) enabler. This allows the platform operator, through a contract with an existing mobile operator, to offer virtual mobile services branded as their own. The platform supplies the enabling systems for the VMO. These applications were not included in the release under test and so did not form part of the scope of this testing.

Finally, *evoca*'s fixed voice services are delivered via an integral softswitch. Applications available include fixed national and international voice resale provided by 'calling card' and 'least cost routing' (LCR), higher margin value add voice services such as conferencing, and personal numbering. The softswitch can support traditional circuit-switched telephony and VoIP. Only the former was included in the scope.

The *evoca* platform is required to be scalable and flexible to enable the platform to generate new applications and services rapidly and cost effectively. It is has therefore been essential to understand the interaction and performance implications of these voice and Internet applications.

6.2.1 The Platform

evoca brings together technical developments and experience produced for BT Group companies and partners around the globe. The platform is constructed from core elements used within operations. This gave a level of assurance of both the functionality and performance of the individual components. The *evoca* platform was integrated from a set of scaled down platform technologies allowing low market entry while maintaining upward scalability.

6.2.2   Key Attributes

The key attributes of *evoca* are summarised below:

- the *evoca* platform exploits reusable components to allow the flexible creation of new services;

- end customers are able to control all aspects of transactions and service via fixed or mobile Web browser interfaces,

- *evoca* is optimised to support multiple vSPs on a single platform;

- the solution supports the connection of multiple wholesale suppliers, with automatic routing between them based on least-cost routing of each call;

- *evoca* supports the use of mobile terminals and Internet access as the enablers for new, innovative services, concentrating on customer profitability and retention;

- services can be targeted at profitable niche markets with easily replicable, scalable platforms.

6.2.3   Key Features

The main features of the *evoca* platform are detailed in this section. Each of these was exercised during the performance testing. The following sections briefly outline the key services and their interaction on the system as a whole.

6.2.3.1   User Registration

The potential end user can register for services via a set of Web screens. During this process the user selects their account name, confirms payment method, and selects the service offering required. Once all required information is collected, the account information is automatically provisioned into the system and appropriate applications. This process has continuous Web interaction.

6.2.3.2   Customer Self-Care (CSC)

Once the end user has registered they can modify their personal details and service selection (addition or cessation of individual services). Through CSC the user can also view their usage and account balance. This process has continuous Web and database interaction.

6.2.3.3 Internet Access Authentication

The *evoca* platform provides RADIUS for authentication of Internet access. RADIUS is automatically provisioned when the users register for this service. RADIUS within the platform has its own database and so real-time Internet access authentication requests do not affect the overall platform.

6.2.3.4 E-mail

The end user can request an e-mail account hosted on the *evoca* platform. The e-mail account can be accessed by PoP3 and/or Web browser. This requires occasional Web/IP interaction.

6.2.3.5 Billing

At the heart of the system is a rating and revenue collection engine. During the testing the rating engine runs as a continuous function and its workload is dependent on the volume of usage data from the individual services. In a real deployment the revenue collection function will be configured to run when the system is at its lowest operational load and as such was not included as part of the peak load testing.

6.2.3.6 Personal Numbering

Web applications allow users to determine the destination number to which calls on their personal number will be sent, or they can be sent to a mail box. Configuration of a required destination can be performed via Web browser or telephony interface. This requires occasional access. The incoming calls are routed based on user pre-configured destination information without any real-time interaction with the 'user'. This service type involves a simple database access during set-up and then requires only a voice path for the duration of the call. The user can access their voice mailbox via either a voice or Web browser.

6.2.3.7 Least Cost Routing

Calls are routed according to a set of rules defined to identify the optimum route based on destination and time of day via simple data processing functions (route costs). This service exercises the system in a similar way to personal numbering, with the addition of some application processing. The changes to routing configuration have negligible impact on system performance.

6.2.3.8   Calling Card

Incoming calls are terminated on the platform. The user is validated by either CLI or a dialogue is conducted to collect card ID and PIN. The call is then routed to an outgoing trunk once the destination information is provided. This service exercises database access as well as the dynamic voice resources during the start of the call and then only requires a voice stream path for the remainder of the call. The user can change their PIN via a Web browser. This has negligible impact on system performance.

6.2.3.9   IVR Applications — Quiz and Vote

Incoming calls are terminated on the platform and a dialogue is conducted with the caller for the duration of the call. The user responds by entering digits to select the required response. There is no call routing function and negligible Web interaction for these services. This service makes constant demands of voice resources and database access. Only one voice stream is required.

6.2.3.10   Conference Call

The conferencing application maintains a list of conference participants enabling automatic notification via e-mail or SMS. Using the Web interface a user can add conference participants' details and assign participants into groups. Conference calls can then be scheduled for a participant group. Only the conference configuration requires Web interaction. At the required time the caller dials in and enters a conference PIN which is validated against the database.

6.3   Approach to Performance Testing *evoca*

As stated, the performance integrity is key to *evoca* because customers demand robust solutions. To ensure that we meet these needs the performance team agreed with the *evoca* designers and developers to adopt a managed-risk-based approach to the performance testing. This involved reviewing the architecture, operation and likely usage of the system to identify critical paths within the system, which allowed concerns over behaviour in certain areas to be scrutinised more closely than others, and the breadth and depth of testing to be designed accordingly. Central to the development aims of *evoca* was the production of a platform that performed and scaled well. Performance testing was therefore planned from the outset with several aims:

- to optimise the platform components;
- to verify the interaction between service elements;
- to confirm that the capacity met design requirements for the required quality of service (QoS) specification;
- to ensure that it withstood sustained loads.

The system architect was particularly keen to verify how work presented to *evoca*'s Web applications affected the performance of the voice services and vice versa, as there was a common database supplying information to each. For example, it was important to ensure that users checking their mailboxes or booking conferences via the Web application did not reduce the ability of the system to support calling card or quiz users, and conversely that heavy use of the quiz function did not slow the Web application response times.

As *evoca* consists of both voice services and Web applications, a team was brought together who had experience of both these environments and a variety of tools were used to exercise the different interfaces.

A full-scale deployment system was shared between the integration and performance testing teams (see Fig 6.2). Working with this environment to obtain meaningful results required close co-operation between the teams and emphasised the benefits of a clearly defined test-plan (driven from the objectives) that enabled test coverage to be tracked and considered judgements to be made on the scope of regression testing in the light of changes to *evoca*.

The whole end-to-end environment was set up in the unique test resource at the Adastral Park Integration Facility (IF), which comprises full-scale instances of both exchange and transmission equipment from vendors used in the BT UK core network.

The system configuration for this exercise did not include the VMO enabling systems, or VoIP telephony. The shared environment also provided some scheduling challenges. A shift rota was devised which saw functional testing taking the main part of the day with the system being handed over to the performance test team mid-afternoon. The performance runs were made during the early evening and over weekends.

A key part of setting up the test environment was populating the databases with a realistic volume of information. The results of the tests could have been misleading if this had not been done, as even a poorly configured database or query can swiftly return a result if it contains only a few records. As part of the stand-alone registration testing, 250 000 end-user accounts were created with test scripts exercising the user interface — this in itself allowed the design team to characterise the system limitations and understand how the platform components would need to scale based on required volumes.

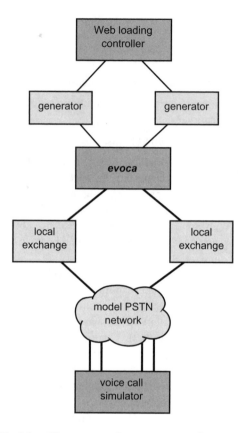

Fig 6.2   The *evoca* performance test environment.

The approach for the first phase was to exercise the individual interfaces in isolation, and within these the separate transactions were individually tested. This allowed the bottle-necks within the components to be identified and rectified as it was much easier to identify 'hot spots' within the system when only one part of the application was being exercised at a time.

6.3.1   Approach to Load Testing the Web Applications

Load on *evoca*'s Web applications was generated using a tool that allows user actions on a Web page to be recorded into a script and then replayed. Scripts can be modified and data built so that each user logs in with separate credentials and search requests, for example, work through a range of information in the database rather than requesting the same information repeatedly, as this may merely exercise the database cache.

Many different scripts can be replayed at the same time to replicate a diverse workload on the system. The rate at which users join and leave and the duration of the test can also be controlled, and, if required, user actions can be synchronised to apply contention to the system. Furthermore the time between page requests (inter-arrival time) can be fixed, varied either formulaically or randomly, and a range of individual browser features can be used, such as cookie and dynamic data handling, browser cache size and refresh policy, browser type, number of simultaneous connections, HTTP version, and connection speed. With careful set-up, this allows the diversity of the Web application user population to be closely replicated. Importantly the response of the server can be checked against expected results to ensure that not only has the Web server responded with a valid response code but, for example, it has logged a user into the right account, not that of someone else.

Each of the business processes described in section 6.2.3 was recorded into a separate script of the Web load test tool. The scripts were tested themselves at low volumes and then they were run individually with varying numbers of virtual users from 100 to 500 accessing the same function. Exercising individual parts of the system in isolation allows the performance of these to be measured and tuned without a mass of other activity clouding the picture.

Once this phase of testing was completed, all of the scripts were brought together into a realistic load simulation scenario, using proportions of users that would be expected once the system was released in a live environment.

6.3.2   Approach to Load Testing the Voice Services

Like the Web performance tests, the voice performance tests were designed to realistically exercise the switch by configuring it in a way that was similar to that which would be offered in a deployment of *evoca*. Using the extensive captive network environment within BT's Integration Facility, a good representation was achieved. Before deployment it is never possible to construct an exact model of the end-to-end call flow mix, as no real historical data exists, and therefore a trade-off between a pure 'stress' test and an assumed level of traffic in a deployment was used; this ensured that the workload was at the challenging end of what was expected.

The resources exercised were:

- switching capability of the voice system, in terms of the internal data path hardware and the controlling database look-ups for the switching;

- the use of the voice resource dual tone multifrequency (DTMF) decode and announcement replay.

In the *evoca* platform the system is being stressed during call processing and set-up, as this requires dynamic system resources. Once a call is set up, only static system resources are required for which there would be no contention and LAN activity would have no impact further on them. The risk of failure of calls once set-up was therefore expected to be minimal. This meant that greater scrutiny of results was required for dynamic resource testing. In addition, it was necessary to use different call generation profiles for each of the two types of testing.

Static resource testing involved setting up a full complement of calls, using an appropriate type mix, and holding the calls for the duration of the test period. Measurements were then taken to check for changes in system resources (memory, database activity, etc) over the test period. Dynamic resource testing involved generating the full complement of incoming calls to *evoca* simultaneously with a relatively short call-duration time. This caused a steep ramp in the number of simultaneous call set-ups required thereby initiating dynamic resource contention within *evoca*. Since all the calls would then expire at the same time and the call generation equipment is set to automatically re-establish calls as they expire, the steep call set-up ramp is repeated. This repeated ramp test is continued throughout the test period during which system response and resource measurements were taken to determine the effect on system performance.

This approach illustrates how and why knowledge of the system architecture and operation is a key element in determining the appropriate form of performance testing.

A C7 signalling system emulator was used to originate and receive up to 240 simultaneous telephony calls through *evoca* in order to load test it. Figure 6.2 illustrated the architecture of the telephony test set-up.

There were three types of call scenario used during the telephony load testing as follows.

- Personal numbering

  When a call arrives at *evoca* from the test platform an outgoing call is made and the two are 'joined' inside the *evoca* platform. Least-cost routing is a similar scenario differing only in the way the destination of the outgoing call is arrived at. These two scenarios behave identically to the test platform and the same script was used to exercise them.

- Calling card

  Calls arriving at *evoca* from the test platform are played messages requesting an account number, PIN and destination number. The tester supplies these in the form of DTMF keyed digits. The *evoca* system makes an outgoing call to the destination number provided and connects the incoming call to it. Dynamic system resources are required during call set-up only.

- Quiz

  Calls arriving at *evoca* from the tester are played a series of messages to which the test platform responds with single DTMF digits. Once all quiz messages are complete the call is cleared by *evoca*. Dynamic system resources are required throughout the call.

6.3.3   Combined Approach

Once the component testing had completed satisfactorily, the second phase of testing began — loading the Web and voice parts of *evoca* simultaneously to determine if this combined load affected the system (the key test for the system architect). Dynamic voice tests (as described in section 6.3.2) were used for these tests as these placed a greater stress on the system and were more likely to reveal bottle-necks than static tests.

6.4   Results

Performance testing commenced at the earliest opportunity and many functional and stability issues were uncovered during the course of the work in addition to bottle-necks and performance issues. As a shared environment was being used, this enabled the development team to make the required changes quickly and so rapid regression testing was possible to confirm the issues had been addressed.

6.4.1   Web Application Results

The Web performance scripts all ran individually without a problem. The act of registration is the most intense on database/Web interaction. The registration has six screens, the last of which initiates the automatic account provision. The registration process was designed to provision 2000 users per hour with a typical duration of 3 minutes. This equated to 100 simultaneous registration attempts. It was also recognised that, due to the way the services may be advertised, a peak registration may exceed this level. The system performed as expected with 100 concurrent user registrations. The system was pushed well beyond design limits and delays were revealed as the loads increased to 500 users all exercising the same piece of functionality at the same time. For example, Fig 6.3 shows the hallmarks of a system reaching exhaustion — as workload increases, the amount of work the system is able to complete decreases. This should be compared to the graphs in Fig 6.5 where the system is subjected to a realistic workload across both Web application and voice interfaces.

These results show that the system slows under massive load but importantly does not compromise the integrity of the system nor produce adverse delays in other applications.

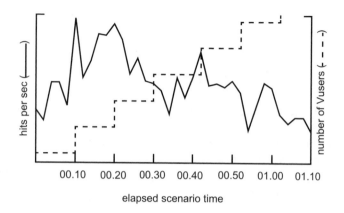

Fig 6.3   The response of *evoca* when stress tested to well beyond its design capacity.

The second phase of Web application performance testing combined the scripts into a realistic mixed workload. The *evoca* platform supported this expected peak of Web users well within the design targets, returning 90% of the pages in less than 1 sec throughout the test run.

6.4.2   Static Voice Results

The low volume (1 call per sec) load testing of the voice system initially revealed a number of hardware faults that required the replacement of components. However, once the system was proved functionally accurate and stable at low loads, it performed faultlessly at loads of about 100 calls per sec with no perceivable degradation in service.

To ensure maximum demand on the key database element, the call hold times for personal numbering and calling card were maintained at the minimum expected in the real-life call scenarios.

The full load of 240 simultaneous calls was apportioned across call scenario types in accordance with data provided by the customer as representative of the expected live call mix.

A continuous load of more than 200 calls was reached within 5 min and this was maintained for a period of 30 min. During the test period the number of simultaneous calls would range between 200 and 240 due to call cycle churn.

During the test period system memory usage, switch route spare capacity, effectiveness of calls (as recorded by *evoca* and the test platform) and database access were monitored to ensure they did not change. In addition, user response times were measured to ensure they met requirements.

6.4.3 Dynamic Voice Results

During this test, dynamic resources were exercised by making calls to all 240 ports simultaneously and measuring the performance from the user perspective. To ensure maximum load was maintained during the test, we needed to make constant demands of the voice resources and database access within *evoca*. When *evoca* experiences a voice resource contention, i.e. when voice resources are instantaneously exhausted, it delays responding to new calls until a resource is freed on another port. This can manifest itself as a small delay in answering the call and subsequent message responses. If the call generation kit waited for an acceptable service delay period to expire before declaring a failure, it would reduce the effective call loading as the voice resource limit would effectively throttle database activity. In order to maintain the demand on *evoca* it was decided to abort a call where the message response was greater than that which would be expected if a resource conflict had not occurred (1 sec). Aborted calls would be reattempted immediately which invokes a database query. It should be noted that this is not a limitation of the testing hardware of the sort discussed in section 6.1.2, but rather an illustration of the fact that testing has to be tailored to the real system's response. Distinguishing between these different situations is a key area of the tester's skill.

Figure 6.4 shows the total number of instantaneous telephony calls in progress throughout the test period. In section 6.3.2 we discussed how a repeated steep ramp nature of call loading is required to ensure the *evoca* platform dynamic resources are exercised continually. This repeated ramp profile can be clearly seen in Fig 6.4 as periodic oscillation of calls in progress throughout the duration of the test.

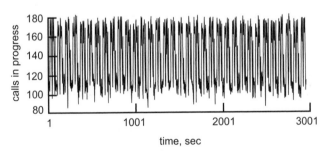

Fig 6.4 Total quiz card calls in progress during test.

6.4.4 Combined Voice and Web Results

The final stage of performance testing was a combination of voice and mixed Web application testing, run to see if the voice application affected the performance of the Web applications and vice versa.

The results are illustrated in Figs 6.5 (a) and (b), and (in contrast to Fig 6.3 which showed system response to a single concentrated application load) show with a realistic mixed workload that as the workload increases so does the amount of work the system is able to complete successfully, i.e. the system is working within its capacity. A small increase in response times for some of the Web pages linked to the database (e.g. register new user) was observed, but these were at a level barely detectable by a human user. Web error rates were unchanged at 0.05%.

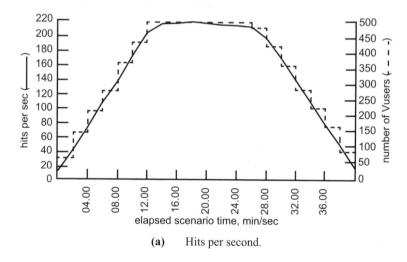

(a) Hits per second.

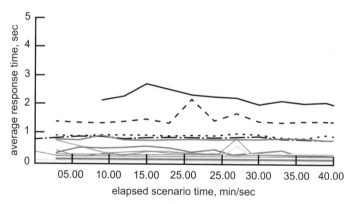

(b) Average transaction response time.

Fig 6.5 Response of *evoca* to a realistic workload across voice and Web applications.

In summary, the low volume (1 call per sec) load testing of the voice system initially revealed hardware faults that required the replacement of components and configuration changes for software optimisation. However, once the system was proved functionally accurate and stable at low loads it performed faultlessly at loads of about 100 calls per sec and 500 simultaneous Web users with no perceivable degradation in service. This highlights the risk of deploying a system based purely on the specification sheets and the benefit of performance testing. Without a true load test, the system could have been deployed with customers discovering the limitations of the system with consequent damage to the reputation of *evoca*. Through the performance testing, the designers and customers of *evoca* can have confidence that the platform is more than capable of meeting its design criteria.

6.5   Summary

The *evoca* platform was designed and built from an integrated set of components. These components were selected because most are used in large-scale deployments by BT within the UK network. Each component was scaled down to give world-class performance at a low market entry level, while maintaining upward scalability. In bringing these components together it was essential to understand how they interacted both functionally and non-functionally. Performance was seen as being a critical success factor, alongside the core functionality and security of the system.

The performance testing carried out demonstrated that the platform was able to meet its capacity targets and exceed the response time expectations. Performance testing and analysis also gave an understanding of the influence of the load mix upon the achievable performance, which will enable customers to deploy *evoca*, appropriately sized to meet their needs. This detailed knowledge of system capacity could not have been achieved by simply bolting the components together 'out of the box'. Performance testing highlighted that there were bottle-necks and identified where these were in the system, in this case being a mixture of hardware and software issues. Once these had been addressed performance testing was able to confirm that they had been rectified and so offer confidence that the product was ready for launch. This demonstrates the place of performance testing alongside other performance management disciplines, and the value of physically undertaking performance tests on an integration of well-designed and implemented components to ensure that they offer the solution expected rather than relying solely on the word of a supplier or a paper performance exercise.

It is also apparent that, to have confidence in the results, the performance tests need to be well designed and carefully executed. They must not only replicate the type of actions that users may undertake, but also the rate at which they use the system, the size of the user population, the configuration of their equipment and the like. Performance testing results in a huge amount of data being collected. Collecting the right sort of data to help answer the key questions driving the performance tests is important but interpreting it correctly is essential, so that the information can be drawn out. However, this includes being able to recognise when the results are unreliable, maybe through injector exhaustion, test script bugs, misconfiguration, and then being able to review the set-up and make the necessary changes. Once the key information has been distilled, this needs to be communicated to the customer for the work and the system implementors so that they rectify any problems. The configuration needs to be carefully managed so that tests can be repeated and the results compared.

In summary, performance testing is a complex activity to get right, but, in an increasingly competitive market, a most valuable assurance.

7

ADAPTIVE NETWORK OVERLOAD CONTROLS

M J Whitehead and P M Williams

7.1　Introduction

Over the last few years, BT has been active in the development of network overload controls. The motivation for this work was, and is, the need for effective, automatic controls to protect a range of service platforms from demand levels that are becoming increasingly volatile and unpredictable. Some of the control techniques developed have been patented [1] and are now implemented and running on several of BT's service platforms (see Chapter 8).

The expertise gained from this work has enabled industry to improve the way overload controls are defined in various UK and international standards. Notable examples are the UK adaptive ISDN user part automatic congestion control (ISUP ACC) [2] and the Asynchronous Transfer Mode (ATM) Forum's Signalling Congestion Control [3].

The ideas discussed in this chapter were first developed for controlling overloads in telephony networks; consequently, much of it draws upon that experience directly. However, many of the basic ideas apply across a wide range of services and technologies with minor variations, and can be fruitfully applied to the newer services and technologies. Examples of this are discussed throughout this chapter, which is structured as follows.

Section 7.2 describes the causes of overloads. This is largely tutorial material and covers, for example, overloads caused by equipment failures, by media-stimulated call surges to specific numbers, and by emergencies.

Section 7.3 discusses the impact of overloads on customer behaviour (specifically customer abandons and reattempts) and on node throughput.

Section 7.4 discusses the variations between overload scenarios, for example variations in the number of traffic sources causing an overload, and variations in the capacity of an overloaded resource.

Section 7.5 derives a generic set of requirements on the behaviour of an overload control from a consideration of the impact of overloads on network resources, and the likely variations between overload scenarios. The requirements discussed include:

- convergence to a state which maximises the throughput of an overloaded resource, subject to keeping its response times short enough to reduce the occurrence of customer abandons;
- achieving such convergence automatically over a realistic range of overload scenarios (including a range of overloaded resource capacities, and numbers of traffic sources at which restriction is active);
- minimising ineffective calls by selectively throttling calls to dialled numbers experiencing high failure rates;
- fairness.

Section 7.6 discusses the principles of good overload control design. The issues discussed include:

- the need for both internal and external overload controls;
- internal overload control issues (workload characteristics and performance requirements);
- external control issues (closed versus open loop, which variable to control, structure of the control, activation and termination);
- control responsiveness and stability;
- an example of good design (the Automatic Call Restriction scheme [1, 4]).

Section 7.7 presents a preferred approach to framing overload controls in standards, and gives some recent examples of good practice. This is an area where BT has been particularly active, specifically in connection with the specification of overload controls by the ITU, Megaco (Internet Engineering Task Force Media Gateway Control working group) and the ATM Forum. It also discusses problems with the way some existing ITU Recommendations have dealt with overload controls.

Section 7.8 discusses new requirements for testing overload controls. It is argued that a new approach is required for testing adaptive overload controls that relies upon both detailed performance modelling and thorough performance measurements.

Section 7.9 describes some potentially useful future developments. Specifically, it discusses overload controls for session initiation protocol (SIP) and hypertext transfer protocol (HTTP), and the control of overloads caused by surges of call releases.

Section 7.10 summarises this chapter, while the Appendix provides technical details of the stability analysis of a class of rate-based overload controls.

7.2   Causes of Overload

In telephone networks, overloads can be caused by the following (either singly or in combination):

- media-stimulated mass calling events — such as televotes, charity appeals, competitions and marketing campaigns;
- emergencies;
- network equipment failures;
- auto-scheduled calling.

In the absence of effective controls, such overloads would threaten the stability of network systems, and cause a severe reduction in successful call completions. Ultimately, systems would fail due to lost tasks, and services would not be available to customers.

7.2.1   Overload Traffic Profiles

Overloads peak at calling rates much greater than the predictable daily profile peak to which the network can be economically dimensioned. Figure 7.1 shows the relative calling rates for a particular service taken over consecutive 15-min periods for a month.

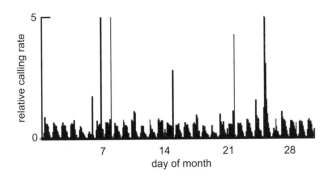

Fig 7.1   Extreme peaks with daily profile.

The overloads can be 4 to 5 times higher than the systematic daily peak quarter hour rate (normalised to one in Fig 7.1), or even worse than that: the vertical axis has been truncated at 5 times the systematic peak quarter hour rate, but in fact peak levels have reached (albeit much less often) as high as 64 times the systematic peak quarter hour rate. Table 7.1 quantifies in how many quarter hours per year the calling rate exceeds increasing multiples of the systematic quarter hour peak.

A network operator could not economically provide sufficient capacity to manage such calling rate peaks. Moreover, in an overload, most service requests cannot be terminated successfully because the terminating lines are busy. Therefore overload controls are needed to manage peaks.

Table 7.1   Extremes of the calling rate distribution.

Calling rate expressed as a multiple of systematic quarter hour peak	Number of quarter hours per year calling rate is exceeded (% of quarter hours p.a.)
2	344 (1%)
4	139 (0.4%)
8	51 (0.1%)
16	22 (0.06%)
32	17 (0.05%)
64	6 (0.02%)

7.2.2   Media-Stimulated Events

Media-stimulated events, such as televotes for TV talent shows, local radio phone-ins, and ticket sales, can generate high calling rates to specific numbers or groups of numbers. In BT's network the frequency of such events is high, currently of the order of several 1000 per month.

The time-dependent calling rate for such events generally has a saw-tooth shape, as shown in Fig 7.2. It has a relatively short rise time — typically of the order of a few tens of seconds to a minute or so — and a slower decay time lasting for several minutes or for up to an hour or more. In some cases the peak calling rate can be sustained for an hour or more. The rate of increase up the ramp can be extremely fast: accelerations of the order of 4000 calls/ sec per sec lasting for 6 sec have been measured in the BT network. In addition, re-stimulation of callers during an event can cause a sequence of sawtooth profiles to be generated.

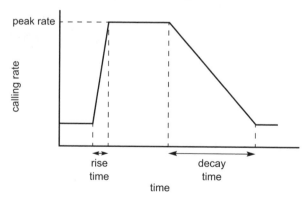

Fig 7.2   Profile of offered rate.

The World Wide Web is also susceptible to media-stimulated high request rates, e.g. to register televotes on-line, or to buy tickets for a sports event.

7.2.3   Emergencies

Natural and man-made disasters can lead to the whole country calling a single contact number advertised via the media, or calling a range of numbers (e.g. bad weather leading to call-out of rescue services, power failures). More often than not the contact number used will be made known to the affected network operator or operators in advance, and they can manually set-up network management controls to manage the load. However, every now and then, the network operator hosting the contact number may not know the number in advance, and network integrity can be threatened by very high calling rates to it; or, the network operator may know the advertised number in advance, but some interconnect routes from other operators to the operator hosting the number may not have suitable (or any) overload controls.

7.2.4   Network Equipment Failures

Significant overloads can be caused by network equipment failures. These cause overload both by reducing the network capacity available to handle demand, and by abruptly terminating existing calls, thereby triggering mass call releases and then a surge of repeat attempts. (The control of mass call releases is discussed in section 7.9.2.) Alternative routing of calls may exacerbate the problem, e.g. if there is a surge of reattempts to an end node which has suffered partial, or complete, loss of capacity. (The best way to integrate alternative routing and overload controls is discussed in section 7.6.3.4.) As failures are by their very nature unpredictable, fast automatic controls are needed to respond to them.

7.2.5   Auto-Scheduled Mass Calling

An example of this phenomenon is where TV set-top boxes simultaneously start (e.g. at 2 a.m.) to dial up a server on the Internet in order to upload or download customer data.

The problem is that there are typically only sufficient terminating lines to handle a small fraction of the calls offered per unit of time, and the instantaneous peak demand may exceed available signalling or processing capacity. One challenging feature of such auto-scheduled events is that the boxes can all follow exactly the same call script, so that — after failure to establish a call attempt — the set-top boxes may then all follow exactly the same repeat sequence, with identical inter-repeat delays.

7.3   Impact of Overloads

7.3.1   Customer Behaviour — Impatience and Repeat Attempts

Customers are intolerant of long response times to their service requests. Typically, if they have to wait more than a few seconds, customers are very likely to abandon their service requests and make repeat requests (reattempts). This is also true of, for example, personal computers executing Internet dial-up scripts, and of TV set-top boxes, both of which have built-in time outs.

Customer impatience, combined with an excessively delayed response to service requests from a network resource, can cause the effective throughput of the resource to collapse to a low level — see Rumsewicz [5] for examples of this phenomenon in Signalling System No 7 signalling networks. This behaviour has been seen in simulations of traffic routes between two telephone switches that do not have effective overload controls. If the switch at one end of the route acts as a bottle-neck limiting the rate at which it can return responses to service requests, then it can be driven into a self-sustaining state in which it has such a large backlog of requests and responses waiting to be processed, that all new requests are abandoned by customers, because the delays exceed what customers will tolerate.

The repeat attempt behaviour of a customer is characterised by specifying:

- the persistence probability, denoted by p_k, that, if the kth service request attempt is rejected, a $(k+1)$th attempt will be made;

- the distribution of the time interval between the kth and $(k+1)$th attempts (the kth inter-repeat delay).

For media-stimulated events, persistence probabilities are usually high ($p_k \approx 80\%$ or more), and inter-repeat delays are usually short — a few seconds on average. In contrast, computer-driven reattempts usually end after some number of repeats (e.g. 10).

The initial attempt (and any subsequent repeat attempts) made by an individual customer to connect to a specific destination number is termed a call intent. It can be shown that, if all call attempts in an intent fail, then the mean number of attempts per intent is given by the series $1 + p_1 + p_1 p_2 + p_1 p_2 p_3 + ...$. If all $p_k = p$ then this sums to $(1-p)^{-1}$. So, for example, an 80% persistence probability would give rise, on average, to 5 attempts per call intent (if all attempts fail). Customer persistence can therefore lead to an 'explosion' of repeat attempts when calls are rejected, resulting in congestion in other parts of the network. It is therefore essential to maintain high effective throughput under overload.

7.3.2   Impact of Overload on Node Throughput

In most network systems call processing is divided among several different sets of processing resources. For example, in telephone switches, it is often divided between peripheral processors which handle sets of traffic routes, and a pool of load-balanced central processors which handle call routing, charging, etc.

Consequently, it is possible that different sets of processing resources could become the bottle-neck as call mixes and the patterns of call flow through the system vary. The system's internal load control therefore needs to reject calls in order to keep response times low and to guarantee correct call handling whatever internal processing resource is overloaded.

For a specific mix of offered demand, the typical throughput curve for a network system is shown in Fig 7.3. The resource could be a telephone switch, a service control point (SCP), a signalling transfer point (STP), an ATM node, a WWW server, etc.

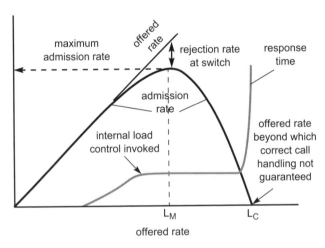

Fig 7.3   Typical throughput curve.

The exact shape of the admission rate curve will depend on which internal processing resources are the bottle-neck for the offered demand mix. However, generally, as the offered request rate increases, there will come a point where the resource invokes its internal overload control to reject some part of the offered load in order to limit the admitted load and consequent delay at the processing resource bottle-neck. At this point therefore, the admitted rate will begin to fall below the offered rate, and eventually reach a maximum admitted rate (since rejecting service requests consumes processing effort) at some definite offered rate (denoted by L_M in Fig 7.3).

Further increasing the offered rate causes the admitted rate to actually fall. Eventually there comes a rate (denoted by L_C service requests/sec) where correct call handling cannot be guaranteed, and all calls are rejected. This can occur, for example, in telephony switches if internal task queues overflow, and the switch may have to restart or roll back to restore itself to a 'sane' state. To restore the resource to normal working, the offered rate will need to be reduced to a level at which the resource is no longer in overload. Thus, as the offered demand is increased to L_C and then decreased, the admitted rate describes a hysteresis curve (not shown in Fig 7.3) that traces the admitted rate curve as the offered rate is increased to L_C, and then returns along the horizontal axis as the load is reduced. Only the application of load restriction at nodes external to the overloaded resource (for example, at the traffic sources causing the overload) can prevent such collapse.

In addition — as is illustrated in Fig 7.3 — the time the target resource takes to respond to a request will start to increase as internal congestion builds up. This response time can be kept to an acceptably low (and roughly constant) level for a certain range of offered rates by means of the target's admission control reducing the rate at which calls are accepted for full processing. Eventually, however, the processing effort of rejecting or discarding calls can itself exhaust the processing capacity available to reject calls (this occurs when the offered load is L_C). At that point response times can start to climb rapidly and uncontrollably, triggering high rates of customer abandons and reattempts.

An additional cause of admitted-rate hysteresis (not shown in Fig 7.3) may exist. It occurs if the switch has an internal load control (see section 7.6.1) which rejects some part of the stream of service requests when the switch is overloaded, and for which the proportion of requests rejected is not a single-valued function of the offered rate over the full range of offered rates. A common cause of such behaviour is the use by a switch's internal overload control of thresholding mechanisms, which stop rapid oscillation between adjacent congestion levels by separating thresholds for upward and downward changes in levels. Then, if different internal congestion levels map to different rejection proportions, the rejection function can take a different value depending on whether the demand is increasing or decreasing.

The effect of such a hysteresis cycle on the control's behaviour (and, in particular, on its long-run behaviour when the offered load is constant) has not been investigated by the authors; and it would be interesting to see how large the cycle can get before control stability is threatened. This chapter assumes that such hysteresis cycles are small enough for their impact on the end-to-end behaviour of a control to be indistinguishable from the impact of the stochastic nature of demand. It is then, arguably, justifiable (as an initial modelling simplification) to assume that such effects are negligible — that is the assumption used in this chapter.

It is important to maximise the target's admitted rate — subject to keeping response times small enough — in order to minimise customer reattempts. This also avoids the sources over-throttling the request streams destined for the target, which would generate unnecessary customer repeat attempts.

The above discussion clearly shows both that automatic external load controls are needed, and also that a good external overload control will adaptively keep the load offered to an overloaded switch close to the rate which maximises its effective throughput — subject to response times being short enough. Poor external load restriction schemes over- or under-restrict or oscillate between over- and under-restriction, and cause a large reduction in effective throughput compared to the target's maximum, leading to a higher chance of call failure and system failure.

7.4   Variations Between Overload Scenarios

Many media-stimulated overload events are known about in advance, and so network management overload controls (such as call gapping) could be applied to try to limit the calling rate to one that will just keep terminating lines busy, or prevent switches getting congested. However, as sections 7.4.1 and 7.4.2 illustrate, the complexity of the network makes setting the overload controls to throttle call streams by the right amount a very difficult task. It is even harder to cope with the many unannounced events that occur. We conclude that overload controls, to be effective, need to be automatic and embedded in the network resources in order to react sufficiently rapidly to sudden surges of demand or changes in network capacity.

7.4.1   Varying Numbers of Overload Sources

For any specific network context, the number of nodes causing a resource to be overloaded can vary widely from one event to another, for example:

- in national circuit-switched networks, the number of neighbouring switches overloading a target switch can range from a few (fewer than 10) up to many (of the order of 100);

- in intelligent networks, SCPs can receive service requests from several hundred service-switching points (SSPs) (see Chapter 8 for details) — in certain national overload scenarios all of the SSPs can cause the SCP to be overloaded, while for a regional overload, the number of sources of the overload could be very small;

- in ATM networks, the originating node selects the entire end-to-end path across a peer group used to attempt to set up a soft permanent virtual circuit (PVC) or a switched virtual circuit (SVC) — consequently, a variable number of originating nodes can cause congestion of the processing resources at any particular node which deals with set-up and clear-down of soft PVCs and SVCs, and therefore the number of overload sources could vary from all the nodes in a single peer group to a small subset of them;

- as an exception to the above examples, a single media gateway is usually controlled by just a single media gateway controller (although H.248 allows a gateway to have more than one controller), and therefore the number of overload sources is typically one.

The conclusion is that in many contexts an effective overload control must be able to cope automatically with variation in the way that traffic is distributed over the overload sources, which may change from one event to another, or even during an event. Such adaptation is crucial to achieving fast and appropriate control of overloads without human intervention.

7.4.2   Varying Target Capacity

In a network, different instances of a particular type of resource may be individually protected from overload by separate instances of the same control. An example would be adaptive ISUP ACC used to protect individual switches. Moreover, the different instances of the resource may have (by design) widely different capacities. ISUP ACC could be used, for example, to protect end nodes (of relatively low capacity) as well as transit nodes (of larger capacity).

Furthermore, an individual target resource's capacity in service requests/sec can vary over time, for example:

- an overloaded resource's capacity (in service requests/sec) can change if the service mix changes — reducing, for example, if it has to carry calls which need more processing;

- a target resource's capacity can change due to partial failure of an internal resource, or due to capacity upgrades, or due to poor scheduling of operating system functions — for example, memory management that fails to respect the response time requirements of user demand (see section 7.6.2.3).

A good control should therefore be able to adjust to cope automatically with all such variations in target capacity.

7.5   Requirements for Overload Controls

The foregoing discussion of the nature of overloads and their impact on network resources serves to motivate the requirements discussed in this section.

A clear specification of how an overload control should behave (i.e. a set of overload control requirements) is essential before deciding how it may best be implemented to achieve that behaviour (the latter aspect is discussed in detail in section 7.3.6).

The derivation of a set of overload control requirements is discussed here in the context of a resource (called the target) which is overloaded by service requests coming solely from a set of resources where restriction is applied (known as 'points of restriction', or 'restriction points'). (This assumes that all overload sources are potential restriction points.) A point of restriction may be located at a neighbouring sending node (as is the case with ISUP ACC), or at the originating node for a service request (as is the case with the private network-network interface (PNNI) protocol when the originating node and the target lie in the same peer group), or possibly somewhere in between the two. The target is assumed to have an internal mechanism to detect if it is in overload and an admission control which decides whether each individual service request is accepted by the target or not. It is further assumed that if the admission control at the target resource decides not to accept a service request, then it deals with it by one of two mechanisms:

- either by sending a response explicitly rejecting the request (together with an indication of overload);

- or by discarding it without notification (see section 7.6.1 for further details).

In some cases the target may not be able to reject requests for service because it has no concept of a call. The H.248 media gateway is an example of this. It essentially acts as a slave to its gateway controller (which does understand a 'call'), and simply follows orders (such as 'play dial tone', 'transcode between a TDM system and an ATM system'). In that case, the target can admit and process them as normal, and also respond with an indication of overload.

The requirements that follow are fundamental in that they imply other, desirable, requirements.

For example, the requirement to maximise effective throughput, subject to bounding response times, implies that a control should not suffer from oscillations with magnitude so large that effective throughput is no longer maximised, or response time bounds are exceeded, or both — see section 7.6.4 for further discussion of stability and oscillations.

As another example, the requirement to maximise effective throughput throughout an overload event, also implies other, subsidiary, requirements, relating to activation and termination of overload control:

- it can be desirable for control not to be activated until the reject rate at an overloaded resource has exceeded some configurable rate, in order to avoid worsening grade of service for short-lived, minor, overloads;

- premature termination of control (and subsequent reactivation of control) can be undesirable for the same reason.

See section 7.6.3.5 for a further discussion of control activation and termination.

7.5.1   Maximise Effective Throughput at Target

The load restriction at the restriction points should automatically and rapidly adapt, throughout the entire duration of an overload event, so as to cause the total load offered to the target resource to converge to the offered rate which maximises the target's admitted rate — subject to the requirement on bounding response times. That is, the control should automatically adjust to a change in the available capacity at an overloaded resource, and to changes in the offered demand at points of restriction. This implies that points of restriction should not restrict subsequent commands relating to service requests that they have already admitted and sent to the target resource. Otherwise, target processing capacity will be wasted on calls which do not complete successfully.

7.5.2   Bounding Target Response Times

It should be possible to configure the overload control at restriction points and target node, so that for the entire duration of an overload event, most response times at the target (overloaded) resource are suitably short. For instance, the 95 percentile of the response time measured in each of a consecutive sequence of 1-sec intervals must not exceed 200 ms.

This requirement is essential to prevent customers abandoning service requests due to long set-up times and subsequently making repeat attempts. It clearly also helps feedback control stability by reducing the delay from the instant of making a change to the restriction level at a restriction point until any consequential change in the reject rate is seen by a restriction point.

It should be possible to configure the overload controls (at both target and restriction points) so that, during the initial transient response of the overload control (i.e. prior to the steady state being reached), the total calling rate offered to the overloaded target should not greatly exceed L_M service requests/sec (see Fig 7.3) measured over any 1-sec period. This ensures that the overload controls should react fast enough to prevent the load offered to the target from getting dangerously close to the rate at which correct call handling cannot be guaranteed and response times explode (see Fig 7.3).

This discussion has only made reference to the response times experienced by demand which directly results from serving calling customers. In reality a system is subject to other workloads which also have response-time requirements, which are in general different (see section 7.6.2.2).

7.5.3   Limiting of Ineffective Calls

Ineffective calls include calls that fail due to busy network terminations, voice-circuit route exhaustion, signalling network congestion, and rejection by system admission control. During overload, ineffective calls waste network capacity and so must be limited. In order to do this, it is essential for an overload control to be able to automatically identify and control the load offered to specific called numbers or destination network addresses attracting a high failure rate due to congestion at a terminating resource, or, in some cases, node identities.

7.5.4   Range of Overload Scenarios

The requirements discussed above should be automatically achievable for any scenario characterised by:

- a step increase in the load offered to a target node from zero to several times the target's calling rate capacity;

- a fast ramp increase in the load offered to a target climbing from zero to several times the target's calling rate capacity over a period of some tens of seconds, followed by a slower ramp down to zero over a period lasting for several average call holding times for the target in question;

- a range of target capacities appropriate to the network;

- a specified range of the number of sources with restriction points overloading the target.

The requirement to define a range of overload scenarios is essential because the previously described performance requirements stipulate how the overload control should behave, but not how it should be implemented. Consequently, overload control implementors need to provide evidence (based on performance modelling or measurements) that their implementations meet the requirements.

The step increase is a likely worst case for any overload control to deal with. The ramp-up and ramp-down calling rate profile serves to test that an overload control implementation can adequately track a varying offered load.

In essence, this requirement implies that the parameters that are used to configure the overload control should work without the need for manual intervention across the required set of overload scenarios, i.e. one set of parameters should work for all scenarios. The more adaptive an overload control is, the more likely it is to satisfy this requirement.

The next chapter discusses in more detail how to set such parameter values for a specific class of intelligent network load-control schemes.

7.5.5   Priorities

It is usually necessary to differentiate between different streams of service requests and enforce differing performance. There are two principle performance measures:

- service request acceptance;

- response time.

Differentiated acceptance of service requests means that some types of request may be rejected, while other types may not, and this may depend upon the state of overload, for example:

- after the initial service request has been accepted, it is usually followed by subsequent demand (both in the forward and backward directions) relating to the initial service request which should not normally be rejected, since it would otherwise lead to inefficient use of resources, incorrect charging if call-clears are discarded, etc;

- emergency calls should not be rejected (as long as the probability of success is not very small) until capacity has been exhausted and all other call types have been rejected — see section 7.6.3.6 for how this may be achieved using priorities.

7.5.6   Fairness with Multiple Sources

When a target is overloaded by more than one source, so that several overload controls (and their associated restrictors) are simultaneously active, it is desirable that — in the steady state — the capacity of the target's congested resources should be divided 'fairly' among the call streams coming from each source.

Fair division of capacity is a surprisingly tricky issue. Several simple definitions are possible. It could mean that the sources see the same probability of being rejected, or it could mean equal division of the target's capacity among the controlled call streams. As will be discussed in section 7.6.5, for overload controls where each restriction point controls the rate at which its calls are rejected, it is not solely the restriction scheme at a restriction point (e.g. proportional discard, leaky bucket) which determines whether the capacity of a congested resource is divided in proportion to the load offered (before restriction) by the sources of the overload. Moreover, such a division of capacity could be undesirable if big streams meet high blocking rates beyond a congested resource, and small streams do not (as could often be the case for focused overloads).

A more systematic discussion of different notions of fairness, and their emergence as properties of the solutions to some specific network optimisation problems, may be found in the literature [6, 7].

7.6   Principles of Good Overload Control Design

This section considers how best to implement overload controls that satisfy the requirements discussed in the preceding sections. It is organised into sub-sections which discuss the following issues:

- why both internal and external overload controls are needed (section 7.6.1);
- factors pertinent to internal overload control design — types of internal workload and their performance requirements (section 7.6.2);
- factors pertinent to external overload control design (section 7.6.3);
- feedback control responsiveness versus stability (section 7.6.4);
- why controlling reject rate is better than controlling reject proportion (section 7.6.5);
- an example of good design (Automatic Call Restriction — section 7.6.6).

7.6.1   Internal and External Control

An overload control has to restrict demand, whether by returning an indication explicitly rejecting it, or discarding it without notifying the sender. In a scheme which uses feedback (i.e. a closed loop control), this is done according to signals or measurements obtained from the overloaded resource. It is useful to distinguish between an internal control, where the system in overload itself does the restriction, and external control, where the system performing the restriction is remote from the overloaded resource.

Shedding demand will always take some non-zero processing effort. This can be made small, but some effort is necessary in order to:

* discriminate so that subsequent demand (after the initial request) is not lost;

* discriminate in favour of priority demand;

* (possibly) obtain additional information such as called address.

It follows that an internal control alone is not sufficient, because as the load increases the capacity available to successfully process demand decreases (section 7.3.2).

On the other hand, control at external control points alone is not reliable because short-term fluctuations in the load sent by external points of restriction, of which there may be many, could mean considerable variation in the load arriving at the overloaded system. If all this has to be processed fully, a large variation in response time would be expected. If the system queueing capacity is exceeded, demand may be lost, and it is better that this is lost in a controlled way than at random. In any case it is not very practical to have only external control because feedback information from the overloaded resource is required to control the level of restriction, and that information may be complex and highly dependent upon the overloaded system. For example, one overloaded system might return information on its varying process occupancies and queue lengths, and another might return information on its varying internal response times. It is not at all obvious how an external restriction point could be designed to handle all such disparate kinds of information. (For the same sort of reason, it is hard to see how to handle the information provided by an overloaded resource that reports how overloaded it is (other than a binary overloaded/not overloaded response) when rejecting service requests — particularly when different systems use different measures of overload level. Fortunately, in the authors' experience, simple binary feedback is sufficient to ensure that the control requirements in section 7.5 can be met.)

We conclude that each system should have its own internal overload control, which can be designed to meet requirements (section 7.5), using whatever measurements are deemed appropriate (e.g. occupancy or response time). Then requests can be either accepted or not, and if not they can be dealt with by either (a) explicitly sending a response rejecting the request, if the protocol allows it, or (b) discarding the service request without notification.

In method (a) the sending system detects request failure by explicit signals, but in method (b) it must be detected implicitly. Implicit detection may be achieved by the sending node starting a timer each time a request is sent, and the protocol returning a response whenever a service request is admitted. The timer is cancelled if the response is received before a time out expires. The duration of this time out should be a little greater than what would be expected for a high load at the receiver. It would also represent the response time budget for this leg of the path (i.e. it must not be too long). The indication of acceptance must be sent by the overloaded node on receipt of the request and not after subsequent communication with other down-stream systems, since the delay is meant to represent the response time of that node and not that of communicating with other systems, which may take much longer.

7.6.2 Internal Workload Characteristics and Performance Requirements

The design of an internal overload control needs to take into account the different types of internal workload and their differing performance requirements. This section discusses these issues.

The workload of a network system does not just consist of that which directly results from serving calling customers, which we term the primary demand, but it can be partitioned into groups which depend upon their characteristics and performance requirements:

- primary demand (section 7.6.2.1);
- secondary demand (section 7.6.2.2);
- kernel workload (section 7.6.2.3).

The two most important performance criteria are described below.

- Response time

  This is the time between two particular events of a process occurring. For example, for primary demand it may be the time between the receipt of a signalling message and the sending of a message in response.

- Workload acceptance

  Some types of demand must be processed and should not, under normal circumstances, be lost. Other types of demand may be totally or partially rejected in order to meet response-time requirements. User data updates fall into the first category, and admitting a call set-up request falls into the second.

Internal overload control should be designed to meet the differing requirements of each workload type. The extent to which this can be achieved is very dependent upon the operating system. We will argue below that only primary demand should be rejected to control load in order that response-time requirements are met. Much has been written about scheduling policies (e.g. Finkel [8]), but these alone do not ensure that response-time requirements are met — they only affect relative response times of the differing workload types. What has been studied less is the simultaneous use of workload shedding and workload scheduling policies, which is a requirement in the context of overload controls.

7.6.2.1   Primary Demand

This is the workload that results directly from responding to the actions of customers, i.e. workload that directly provides the services that are the ultimate purpose of the system; for example, this would include control of call set-up.

Primary demand is characterised by the initial demand, which if accepted leads to subsequent or follow-on demand for call control. There may be distinct phases to the subsequent demand, e.g. mid-call processing and call clear-down.

Each phase of processing primary demand has associated performance requirements. In the case of call set-up, there should be an upper bound on the set-up time in order to discourage the caller from abandoning the call. If the caller does abandon a call in set-up, then in many cases the system continues to process the call even though it is useless work, which results in inefficient utilisation of the system.

Initial workload may be rejected in order to limit the system utilisation and meet response time requirements, but follow-on workload should not normally be lost.

The use of signalling-message-priority mechanisms normally leads to better throughput of completed calls. However, simple priority schemes, where all queued high priority signalling messages are served before low priority signalling messages, can under certain conditions [9] produce undesirable 'synchronisation' of signalling message priorities giving rise to severe oscillation in the call throughput. More subtle priority management methods are required to avoid this.

7.6.2.2   Secondary Demand, Including Management of System Resources and User Data

There are many facilities that are necessary to support the processing of primary demand which generate additional workload. Some of these originate from the system itself, whereas others originate from users which are not those that generate the primary demand. These facilities include:

- generation of alarms/events;
- generation of system performance measurements;
- management of system resources (e.g. running overload controls);
- monitoring and control of user processes (applications);

- provisioning of user data.

There is consequently a large variation in the workload characteristics of secondary demand.

Generally this workload should not be shed, and it is, in general, less time-critical than primary demand except for any alarms that require a fast automatic response.

7.6.2.3   Kernel Workload

The operating system itself generates workload due to managing itself and providing many services to the other (user) processes running on the system.

Such functions are highly dependent upon the operating system. For Unix, the execution context refers to the mode in which kernel functions are executed [10]. These modes are either process context or system context. In process context, the kernel acts on behalf of the current process (e.g. while executing a system call). System-wide tasks are not performed on behalf of a particular process, and hence are handled in system context (also called interrupt context). It follows that the workload due to kernel functions can correspondingly be divided into system and process workload. The main kernel tasks executed in system context are process scheduling and memory management.

Clearly system workload is important, so it should not be shed, and may have stringent response time requirements (e.g. scheduling) — but it still needs to respect the response time requirements of primary demand.

7.6.3   External Overload Control Structure

7.6.3.1   Closed Loop and Open Loop Overload Controls

Closed loop overload controls use information about the state of the overloaded resource (or resources) in order to adapt the load admitted over time. By choosing an appropriate variable to control and designing the control in the right way, it is then possible for the overload control to adapt so that the admitted load meets the requirements presented in section 7.5. Open loop overload controls do not have such feedback and therefore will not, in general, meet the requirements under varying loads (e.g. varying request rates, call complexity, number of sources) and are therefore generally not adequate.

However, because it takes some time for information to travel around the control loop, feedback controls are vulnerable to rapid changes in the offered load; and, even for a constant offered load, large enough delays around the feedback loop can cause instability. A combination of open loop overload controls which are active when load is extremely variable, and closed loop overload controls otherwise, can be used to overcome the problems caused by rapid changes in load.

7.6.3.2 The Controlled Variable

It was concluded in the preceding discussion that each internal overload control should have a means of shedding requests. Hence it is natural for external restriction to use, as a control variable, a measure of overload based upon registering rejected requests.

Arguably the simplest and most effective measure is request failure (overflow) rate at an overloaded resource. This has the following properties:

- it only requires indications of failed calls — all network resources provide an indication when a request attempt fails (or it can be inferred by using a time out) and thus it is easy to construct, from such indications, a measure of the overflow (failure or reject) rate (this external view of the resource as a 'black box' is convenient because it requires no further development of the resource);

- as the offered load increases and a system approaches overload, i.e. the response time gets close to the maximum requirement, the internal overload control begins to shed requests, and the request failure rate climbs quickly — at such levels the rejected rate has a high rate of change as a function of offered traffic, and this rate of change soon approaches the rate of increase of the offered traffic, and so it is a good measure with which to control offered load.

Furthermore, if a threshold is placed on the failure rate, and the admitted rate is increased whenever the failure rate drops below this threshold, or is decreased whenever the failure rate rises above this threshold, then:

- it provides direct control of the level of ineffective demand in the network, and thus the workload that is wasted in the network;

- it also maximises effective traffic subject to limiting response times;

- it can be shown that if each of several sending nodes has its own goal request failure rate, then there is a single, predictable, steady outcome in terms of the admitted rate at each restriction point — see the Appendix at the end of the chapter for the technicalities.

Contrast this use of request failure rate with the use of the occupancy of the overloaded resource, whose rate of change at high load is small since the occupancy of an overloaded resource will be near its maximum value. This fact means that large changes in the offered load will produce small changes in the measured occupancy, and that is unlikely to lead to a satisfactory control mechanism.

As an example, consider a very simple resource: a group of terminating telephone lines. A comparison of overflow traffic per line and occupancy (Fig 7.4) for different line group sizes shows that overflow is a superior control variable at high occupancies. Here the rate of change of occupancy, as a function of the offered load, is small, which means that the inverse function, which would be used to derive what the offered load is in terms of the measured occupancy, is not very precise. Note that, as the offered traffic increases, so the occupancy of a group of lines

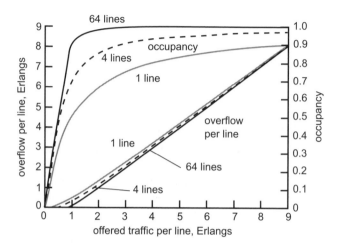

Fig 7.4   Occupancy and overflow as a function of offered traffic.

approaches 1 and the overflow traffic increases quickly. Thereafter as the offered traffic continues to increase, the overflow rate approaches the asymptote where any additional load will be shed.

Of course, the overflow traffic in Erlangs cannot be measured directly because it depends upon the amount of work placed on the overloaded resource by each access attempt, which in the case of a group of telephone lines is defined by the holding time. However, it turns out that such dependence can safely be ignored if an overflow rate limit is chosen which ensures that the lower limit of occupancy is exceeded for overload scenarios in which the calls have the shortest average holding time. The occupancy will also depend upon the number of lines in the line group, but it can be shown that for given overflow traffic, the occupancy increases as a function of the number of lines (Fig 7.5). Thus if a level of overflow traffic is chosen such that the lower limit of occupancy is satisfied for 1 line, it will then be satisfied for all line group sizes. It will also be satisfied for all expected traffic if the overflow traffic limit is chosen for the shortest average holding time. For example, if it is desired that the occupancy should always be above 95% when in overload and the minimum call holding time is 10 seconds, then the overflow traffic limit is approximately 18 Erlangs (Fig 7.5), and the overflow rate limit should be set at 18 Erlangs/10 sec mean call hold time = 1.8 rejects per sec.

We have shown that the request failure rate is a good control variable for external restrictive control, and arguably the simplest. It is possible to use other functions of the call failure counts directly, e.g. the proportion of requests which are rejected. However, it can be shown that this suffers from not always having a unique and stable solution to which the control converges (see section 7.6.5).

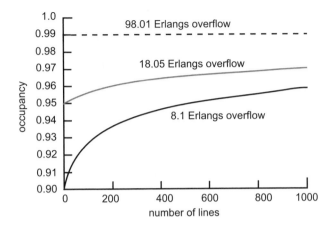

Fig 7.5 Occupancy as a function of number of lines for specified overflow traffic.

7.6.3.3 Functional Components

Feedback overload controls which are based upon the use of detection of overload by monitoring failed requests can be viewed as having the following functional components (see Fig 7.6).

- Detection and monitoring of overload (denoted D in Fig 7.6)

    The inputs for this are indications of resource access failures (rejections, etc) which may include an identity such as node identity or address digits, and the output of which is used to update the level of restriction on such specified identities.

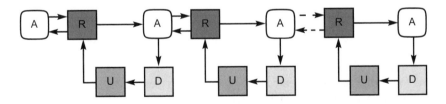

A - admission and routing control
D - detection and monitoring
U - updating restriction parameters
R - restriction

Fig 7.6 The basic component processes of adaptive overload control based on feedback of call rejects.

- Updating of restriction parameters (denoted U in Fig 7.6)

  The input for this is the output from the function D, and the output is the current level of restriction to be applied as defined by the relevant restriction parameters together with the identity to which it should be applied.

- Restriction of demand (denoted R in Fig 7.6).

  Access attempts are admitted or rejected according to the current state of the restriction process, for which several algorithms are possible (see section 7.6.3.6). The level of restriction applied is determined by the most recent indications received from the restriction updating process U. There can optionally be multiple instances of R and U based on, for example, the called-address digits of calls which fail.

The functional configuration of these component processes and the information flows between them for a single stream of demand are indicated in Fig 7.6. Call set-up is from left to right. The mapping on to network systems is deliberately not presented. A further admission and routing control component (denoted A) is added to represent where access to network resources is controlled, which includes access to the bearer and signalling network and functions such as re-routing under congestion, load sharing, and advanced service features. During a telephony focused overload, overload is most likely to first manifest itself by a high rate at which calls fail to access the destination lines on which calls are terminating. This overflow traffic will be detected by the detection and monitoring process at the node hosting the lines. At points in the network between the source of demand and where it terminates there will be control of access to other network resources such as voice trunks, ATM links, signalling resources, etc. Failure at such points means that all resources determined by the routing have been attempted, and that access has been blocked to all of them. The failure is registered by the associated detection and monitoring process.

7.6.3.4   *Staged Overload Control and Interaction with Routing*

Each sequence of component subgroups R, A, D, U of Fig 7.6 forms an overload control loop which can be repeated many times. This is termed staged restriction. In a truly focused overload this will mean that restrictive control will be applied progressively further and further away from the focus (termination) of the overload as the severity of the overload increases. This has the following advantages:

- lower control loop round-trip times (more stable control);

- avoidance of problems with translation of called number/address (a control message sent back to an originating node could otherwise refer to an identity which is meaningless to that node);

- robustness to network failures, i.e. failures will not result in loss of a control loop since control communication can always start adjacent to the point of failure.

It can be seen from Fig 7.6 that another advantage of applying overload control in stages is that the overload control does not interfere in any unusual way with alternative routing or load balancing applied at the function *A*, i.e. a request rejected by call restriction will be subject to the usual routing policy for failed requests.

This is arguably the ideal way to handle a surge of reattempts following a network failure that prematurely terminates established calls. It means that call streams may, for example, freely route around failed nodes if alternative routes with idle capacity exist. But, if such re-routing causes a downstream node to become overloaded, then a restrictor will be activated, forcing more of the call stream to try other alternatives. A call that overflows all available alternative routes, is then counted as a failure by the local detection and monitoring function, and a high enough failure rate will activate control upstream. This mechanism, invoked across the network as required, will bound the total rate at which calls fail due to node congestion or no free circuits.

The use of alternative routing not integrated with a staged overload control, should be regarded with caution, because of the danger of spreading processor overload.

7.6.3.5   Activation/Termination of Overload Control

Detection and activation of overload control must be rapid because of the severe nature of some overload onset profiles (as illustrated in Fig 7.2).

It should be possible to configure the initial level of restriction at all sources to a (suitably severe) value to allow network operators to cope with their envisaged worst-case step increases in demand, or step changes in target node capacity. Different operators may wish to initialise the restriction to different levels, depending upon their networks and requirements. This will require co-operation between network operators when the demand causing an overload crosses network boundaries. However, this configuration should only need to be done once provided that the control is powerful enough to automatically adapt to cope with the anticipated variations between overload scenarios.

It can be desirable for overload control not to be initiated until the node hosting the overload control detects that its calls are being rejected at a rate greater than an operator-configurable rate. This provides an adjustable trade-off between speed of control activation and its impact on grade of service (GoS).

Termination of overload control also needs to be designed carefully. In particular, control should not terminate prematurely, resulting in a new surge of traffic which can itself then cause reactivation of overload control. Once this cycle is started it may be difficult to stop because of the effect of repeat attempts and it may result in low throughput.

One way to prevent premature termination of an overload control is to require that the control of load towards a target should be terminated at a source when both the rate at which the target rejects calls it receives from the source, and the rate at which the source rejects calls destined for the target, have been close to zero for a sufficiently long period to indicate that the overload has abated.

Another possible way to prevent premature termination is to require that control behaves as follows. When the restriction level has adapted down to a minimum level of duration, a termination pending timer should be started. If the restriction level needs to increase before the timer expires, then the restriction level should adapt up from its minimum level and the termination pending timer should be cancelled. Otherwise, when the timer expires, overload control ceases.

7.6.3.6   Restriction Algorithm and Properties

The restriction of demand (denoted by R in Fig 7.6) is accomplished by a restriction algorithm (or restrictor for short). A restrictor is a mechanism for thinning a call stream. Examples of restrictors are given below.

- Proportional rejection

  Reject each new call with some definite probability, p (the restriction level). The admitted rate is then $\lambda(1-p)$ when λ is the call arrival rate at the restrictor.

- Call gapping (Crawford algorithm)

  Upon admitting a call, start a timer of duration τ sec. The restriction level is τ, the gap interval. Reject all subsequent calls which arrive at the restrictor before the timer expires. If the arrival stream is Poisson with rate λ calls per sec, the admitted rate is $(\tau + \lambda^{-1})^{-1}$ calls per second.

- Continuously leaking bucket

  This is a count that continuously decreases at rate r (subject to not falling below 0). When a call arrives, if the count is less than the bucket depth (the maximum value of the bucket count), then the call is admitted and the count is incremented by 1; otherwise the call is rejected and the count is not incremented. The maximum sustained admitted rate is r.

A continuously leaking bucket with bucket depth $= 1$ is the same as call gapping using the Crawford algorithm with gap interval r^{-1}.

When choosing a candidate method for standardisation, there are at least three points to consider:

- effectiveness at meeting the overload control requirements;

- ease of implementation and processing overhead;

- fairness.

Let us compare the proportional rejection scheme with the commonly implemented leaky bucket scheme.

Firstly, we look at the effectiveness of meeting the overload control requirements. Proportional discard has the significant disadvantage that it does not bound the rate at which it admits calls. That means that the overload control has to be able to adapt the discard proportion quickly to keep track of rapid changes in the offered load and ensure that response times at an overloaded resource are bounded.

By contrast, the use of a leaky bucket restrictor has the advantage that it does bound the rate at which it admits calls. As a consequence, the admitted rate is largely insensitive to changes in the offered rate if the offered rate exceeds the bucket's maximum admitted rate. That, in turn, minimises the amount of adaptation of the restriction level required by an overload control, and makes it far easier to bound response times when the offered demand is varying.

Next, ease of implementation and processing overhead — both schemes are implemented by most switch suppliers, and both are well defined and simple to implement.

Finally, fairness — as is discussed in section 7.6.5, it is not solely the restriction scheme which determines whether the capacity of a congested resource is divided in proportion to the load offered (before restriction) by the sources of the overload. Moreover, such a division of capacity will be undesirable if big streams meet high blocking rates beyond the overloaded resource, and small streams do not (as could often be the case for focused overloads).

For these reasons, the use of leaky bucket restrictors is strongly preferred to proportional rejection.

It may be required to incorporate call priorities into restrictors. This can be done in several ways, for example, by stipulating that restriction is applied firstly to low-priority calls and only extended to high-priority calls in a specific overload scenario if necessary. Call restriction needs to be modified to correctly reflect the use of call priorities, and there are simple ways to do that (see, for example, section 2.6 of ITU-T Recommendation Q.714 [12].

7.6.4   Responsiveness and Control Stability

Feedback overload controls have two opposing objectives:

- responsiveness to changing load;

- stability when the load is constant.

The behaviour is a compromise between these two. Not responding quickly enough under an increasing load can lead to under-control and hence excessive response times or lost signalling messages. On the other hand a control which is too sensitive may cause oscillatory throughput even when the offered load is constant, which can mean a lower call completion rate, or violation of response time requirements at an overloaded system. Performance modelling is required to prevent oscillations of such magnitude that throughput or response times suffer. Small magnitude oscillations, which do not breach those requirements, and which do not cause undesirable interactions with downstream systems, are not a problem.

Modelling work has shown that control instability can be caused by sluggish estimation of:

- call reject rates at restriction points;

- resource load level (e.g. processor occupancy) at an overloaded system;

see also Chapter 4.

Informally, a set of interacting feedback controls is globally stable if, for a constant load and whatever the initial state of the system — it always converges close to a unique state as time unfolds. Such behaviour is clearly preferable to alternatives such as regular (or worse still, chaotic) oscillation.

One initial way to establish the stability of a set of interacting feedback controls is to assume that their evolution over time can be approximated by a set of ordinary differential equations (ODEs), and then try to prove that there is a unique solution which is locally or globally stable using any suitable mathematical technique. This kind of analysis usually ignores stochastic effects (because it deals with ordinary differential equations) and time lags. This is acceptable provided it is just the first step in the design of the overload controls. Subsequently, more detailed modelling (usually by simulation) is required which takes those aspects into account. The attraction is that we can actually prove that a system is stable (albeit given certain assumptions) — which is something that no amount of simulation can do.

The principal mathematical method for establishing the stability of a dynamical system described by a set of ordinary differential equations is the use of Lyapunov functions [12, 13]. It is capable of proving that a system is globally stable, and has been applied to the Internet [7], to ISUP ACC [14], and to ATM networks. It may also provide information on speed of convergence [13].

To illustrate this method of stability analysis, consider the class of feedback controls shown in Fig 7.7. The class is characterised by an admission control A at the target node, and individual instances of D, U and R at each source node — see section 7.6.3.3 for their definition. The essential feature of this class of overload control is that each source node continuously measures the rate at which its calls to the target are rejected due to target overload. Based on that data, it adaptively adjusts the rate it admits calls (and hence offers them to the target) so that its measured reject rate converges to a local reference reject rate.

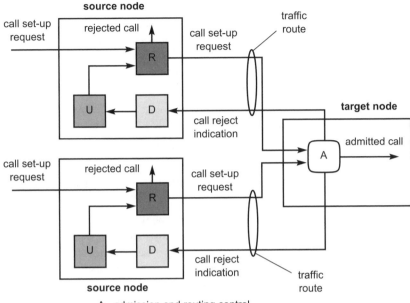

A - admission and routing control
D - detection and monitoring
U - updating restriction parameters
R - restriction

Fig 7.7 Class of feedback control schemes.

Now, this is clearly an incomplete specification, and could not be implemented without defining in detail exactly how the three processes *D*, *U* and ' actually work. There are public domain examples [4, 15] that are covered by a patent (in the USA [16] and in Europe [1]).

Despite this incompleteness, there is sufficient information to analyse two key aspects of the behaviour of this class of overload controls — steady state behaviour and conditions that ensure convergence to the steady state.

In the Appendix to this chapter it is shown that the steady state of this class of overload control (if it is achieved):

- is unique;

- maximises the effective throughput of the target node, whatever its capacity may be, and however many sources there are, provided that the sum of the individual detection and monitoring process reference reject rates at the sources is small compared to the capacity of the target node.

Also in the Appendix at the end of this chapter, it is shown that convergence to the unique steady state is always achieved given some mild constraints on the behaviour of the overload control instance at each restriction point. This kind of convergence result can be absolutely crucial when drafting text for a national or international standard. If the various interested parties cannot agree upon a specific detailed overload control design (which is often the case), then it is of great practical importance to have some assurance that different suppliers' overload control implementations will work together adequately provided that they all belong to the same class of overload controls. This was the approach successfully advocated by the authors:

- within the UK when drafting issue 3.0 of the UK national version of ISUP [2];
- when drafting the UK variant of BICC (bearer-independent call control) ACC [17];
- internationally at the ATM Forum when drafting the ATM Signalling Congestion Control [3].

The essential idea of a Lyapunov function can be illustrated for a simple instance of this class of overload control. The instance consists of two sources jointly overloading a target node with capacity L_M calls per sec. Source node i adaptively adjusts the rate (denoted by $\gamma_i(t)$) at which it sends call requests to the target so that the rate at which its calls are rejected by the target node is driven to a reference rate of r_i rejects/sec. In this example, the state of the control system is the vector of admitted rates $(\gamma_1(t), \gamma_2(t))$. The rate-based overload control is assumed to adapt the admitted rate of node i according to the following equation:

$$\frac{d}{dt}\gamma_i(t) = \kappa\left(r_i - \gamma_i(t)B\left(\sum_j \gamma_j(t)\right)\right)$$

where κ is a positive constant, and the target node blocks arriving calls with probability $B(\Gamma)$ when it is offered calls at a rate of Γ calls/sec. Note that κ influences only the rate of convergence, not the steady state. For simplicity, we assume that the admission rate curve of the target node is the following special case of the general switch throughput (Fig 7.3). It admits all calls when the offered rate does not exceed L_M, and limits the admitted rate to L_M otherwise. The fact that the admitted rate curve remains horizontal in this example means the node can reject calls with zero processing effort (see Fig 7.8).

For this case, $B(\Gamma)$ is given by:

$$B(\Gamma) = \begin{cases} 0 & \text{if } \Gamma \le L_M \\ 1 - \dfrac{L_M}{\Gamma} & \text{if } \Gamma > L_M \end{cases}$$

where L_M is the maximum admission rate of the target node. Also, it follows, from the analysis in this chapter's Appendix, that — at the steady state — the admitted rate from source node i is given by:

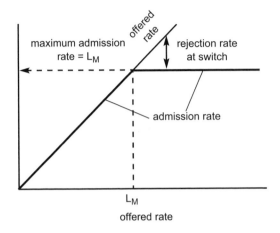

Fig 7.8   A simple instance of a throughput curve at the target node.

$$\gamma_i = \frac{r_i}{\sum_j r_j}\left(L_M + \sum_j r_j\right)$$

Figure 7.9 shows the direction field of these differential equations for the case $L_M = 100$, and $r_1 = r_2 = 2$. Each arrow shows the direction of the solution curve passing through its base. Superimposed upon the field are four specific solution curves, starting from states (80,0), (0,0), (120,30) and (110,110). Observe that they all tend to the (unique) equilibrium point $(\gamma_1, \gamma_2) = (\frac{2}{4}(100 + 4), \frac{2}{4}(100 + 4)) = (52, 52)$.

Informally, a Lyapunov function for a set of differential equations is any scalar function of the system state which has the following properties:

- it has continuous first partial derivatives with respect to the state;
- it takes the value 0 at an equilibrium point;
- it is positive at all points sufficiently close to that equilibrium point;
- along all solution curves near the equilibrium point its derivative with respect to time is less than 0.

It can then be shown that, near the equilibrium point, all solution curves will converge to the unique state which minimises the Lyapunov function. Intuitively, along any solution curve the system state moves downhill on the Lyapunov function and can only come to rest at the point where the function is locally minimised. Figure 7.10 shows the contours of a Lyapunov function for this system, given by equation (7A.11) in the Appendix.

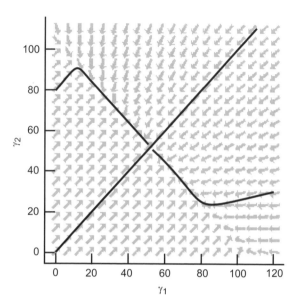

Fig 7.9   Direction field of differential equations.

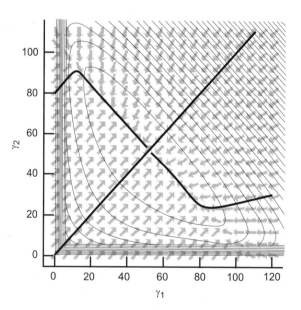

Fig 7.10   Contours of Lyapunov function superimposed on direction field.

The power of Lyapunov's method comes from the fact that one does not need to be able to solve the set of differential equations to apply it. One just needs to be able to form the rate of change of the Lyapunov function along any solution curve; and that is obtained by taking the partial derivatives of the Lyapunov function and combining them with the known form of the right-hand side of the differential equations, as explained in the Appendix at the end of this chapter.

It should be noted that the analysis in the Appendix does not take into account the effects of, for example, signalling and nodal processing delays, or the random nature of the offered load. Consequently the constraints must be regarded as sufficient for convergence given that the use of ODEs to model the overload control is a valid approximation, but perhaps not generally sufficient.

7.6.5 Reject Rate or Reject Proportion?

The choice of which controlled variable to use (reject rate or reject proportion) has important implications for both the fairness of the overload control and its stability.

Consider first the fairness issue.

The Appendix shows that — for the class of overload control defined in section 7.6.4 — in the steady state, each source switch gets a share of the overloaded node's effective capacity which is proportional to the source's reference reject rate. Thus, if the source nodes have equal reference rates, the target node's capacity is divided equally between them.

This fact is true irrespective of the specific restriction method used at the sources by the overload control, because the analysis in the Appendix simply models the effect of a restriction scheme by its instantaneous admitted rate. So, in particular, it is true even if the sources use proportional rejection. Note the important implication: if one decides to control reject rates, then the use of proportional rejection does not lead to a division of the overloaded resource's processing capacity between the controlled sources in proportion to the loads offered to them before restriction. Such fair division of congested capacity is not a feature of controls which seek to equalise reject rates.

Thus, if it is required that the target's capacity be divided among the controlled sources in proportion to their offered load (before restriction), then the variable controlled by the feedback control at each source cannot be its reject rate — unless the overload control is (substantially) modified so that each source dynamically adjusts its reference reject rate to be proportional to its offered calling rate before restriction.

Consider next the stability problems that controlling reject proportions can cause.

We change the simple example discussed in section 7.6.4 so that each source adapts its admitted rate so as to drive the proportion of its calls that are rejected by the overloaded node to a specific reference proportion g. The differential equation governing the admitted rate of node i is then:

$$\frac{d}{dt}\gamma_i(t) = \kappa'\!\left(g - B\!\left(\sum_j \gamma_j(t)\right)\right)$$

where κ' is a positive constant.

The direction field changes to that shown in Fig 7.11 for the case $L_M = 100$, and $g = 0.025$.

In contrast to Fig 7.9, it is clear that this reject-ratio-based overload control has no single equilibrium point. Instead the direction field moves any initial state along a line of slope 45 degrees with the horizontal axis until it reaches the 'capacity line':

$$\gamma_1 + \gamma_2 = \frac{L_M}{1-g} \approx 102.6$$

where it rests. Any point on that line is an equilibrium point for the overload control. Unfortunately, none of the equilibrium points on the line is stable: if the state is given a small push along the capacity line then it will not be driven back to its original state by the system dynamics. This has two undesirable consequences:

- the point to which the overload control converges depends upon the initial admitted rate at each source when the control is activated — that could make it unpredictable in a mixed vendor network;

- even if the sources have the same initial admitted rates, small random nudges of the system state will cause it to execute a random walk along the capacity line, potentially taking it far from any sensible operating point.

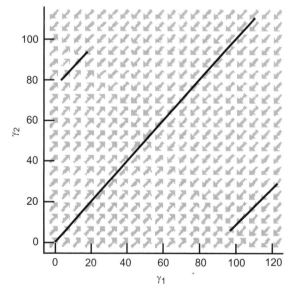

Fig 7.11   Example of a direction field when the controlled variable is the reject proportion.

These problems can only be cured by changing the direction field to one which drives each initial state to a unique and stable equilibrium point. There are ways to achieve this, but only by imposing further conditions on the way restriction levels are updated. That is both unnecessary for adequate overload control (i.e. controlling reject rates is preferable) and likely to be unattractive to suppliers who have to implement a standardised overload control.

7.6.6   Automatic (Dynamic) Call Restriction

An adaptive overload control that automatically controls focused overloads in any telephony network, including intelligent networks, has been designed and patented [16]. It satisfies the design principles presented in this section and the requirements presented in section 7.5. Its application to intelligent networks is covered in Chapter 8 and further details can be found in Williams [4]. Briefly, its detection and monitoring function (see section 7.6.3.3) uses leaky buckets:

- for quick detection of the onset of overload (because the time it takes for the bucket to fill to the onset threshold [4] is inversely proportional to the call failure rate);

- to monitor call failure rates due to overload.

Its restriction updating function (see section 7.6.3.3) is able to make:

- small changes to the restriction level when the measured reject rate is close to the target reject rate;

- progressively larger changes to the restriction level when the reject rate is far from the target rate.

7.7   Overload Controls in Standards

7.7.1   Examples of the Preferred Approach

The preferred approach to framing an overload control standard is that it should, as a minimum, completely specify the overload control's behaviour. That is, the standard should pin down the control's required behaviour as discussed in section 7.5, and possibly including some of the preferred implementation techniques discussed in section 7.6.

Two key benefits flow from this approach:

- the control can be implemented in any way a node supplier wishes provided only that it meets the requirements on behaviour;

- network operators are assured that the behaviour of a compliant implementation of the overload control will be acceptable to them.

Such an approach is now beginning to find favour in national and international standards organisations. Recent examples include the following.

- UK adaptive ISUP ACC [2]

  This issue of the UK ISUP standard has text specifying the required behaviour of adaptive automatic congestion control (ACC) in order to protect a node from call surges coming from neighbouring nodes across ISUP routes. The text follows the approach described in section 7.5.

- UK adaptive BICC ACC [17]

  The latest UK draft BICC standard has text specifying the required behaviour of adaptive ACC in order to protect nodes providing the call service function (CSF) from calling rate surges. The text closely follows the approach described in section 7.5.

- ATM Signalling Congestion Control [3]

  This new overload control standard is designed to protect ATM nodes from SVC or soft-PVC calling rate surges. When an originating node (or, more generally, any node which changes or creates the designated transit list carried by the PNNI set-up message) receives a release message indicating a set-up has encountered signalling congestion, it should use it (a) to drive control adaptation, and (b) may optionally offer the call to an alternative end-to-end route if a suitable one is available. The standard stipulates how the control should behave (along the lines described in section 7.5).

Also, two overload control standards are being progressed at the ITU-T, specifically:

- adaptive ISUP ACC is being progressed in questions 6 and 8 of Study Group 2 [18], with the aim of incorporating a specification of the behaviour of adaptive ISUP ACC in the ITU-T Network Management Recommendation E.412;

- an H.248 overload control package to protect a media gateway from processing overload is being progressed in question 3 of Study Group 16 [19] — it is designed as a feedback control capable of adaptively throttling fresh demand offered from one or more media gateway controllers to a single overloaded media gateway.

In addition to the above approach, useful material is to be found in ITU-T Recommendation E.744 [20]. It discusses (among other things) overload control requirements and implementation considerations for MTP, SCCP and IN overload controls.

7.7.2   Problems in Current Standards

Many international standards are (or until recently were) defective in the way they define(d) controls to protect nodes from processing overload.

The authors' experience furnishes the following examples.

- Incomplete specification of control behaviour

There may be little or no attempt to define how a control should behave in performance terms (contrary to section 7.5). For example, actions upon receipt of an overload indication are often left as 'implementation dependent'. Examples are to be found in ISUP ACC as defined in Q.764 and related ITU-T Recommendations (e.g. E.412), in BICC ACC [21], in the H.248 Media Gateway Resource Congestion Handling package [22], and (until recently) in the absence of any PNNI overload controls to protect ATM nodes against surges of SVC set-up requests or soft-PVC set-up requests.

- Apparently no performance modelling of controls

Few overload control standards cite performance studies as evidence that a control works well. In the authors' experience, the absence of such evidence often means the control turns out not to perform adequately (i.e. it fails to meet the generic requirements discussed in section 7.5). Examples of poor performance are provided by the overload control mechanisms specified by Interconnect User Part (IUP) [23] and ISUP [24].

IUP uses a restriction mechanism known as temporary trunk blocking (TTB). Recent simulation work has shown that TTB may not scale adequately to protect high-capacity switches.

Work at the Public Network Operators Interconnect Standards Committee (PNO-ISC) has shown (see BSI PD6645:2002 [23]) that:

> 'The fundamental problems with the ITU-T ISUP ACC procedure are that it is not a feedback control, is coarse-grained and incomplete. Implementations based on it are unlikely to work adequately.'

The cited study revealed that ACC can cause control actions to synchronise across restriction points, so that they turn on and turn off together, failing to converge.

- Poor choice of restriction algorithm

Many standardised controls recommend the use of the proportional rejection mechanism without discussion of better alternatives (such as the leaky bucket). Telephone User Part (TUP) [25], Signalling Connection Control Part (SCCP) [11], and H.248 Annex M.2 provide examples of this.

- Use of one restriction level per overload level

Some standardised controls define a number of levels of overload that an overloaded node can experience, and then recommend that the call restriction mechanism should employ an equal number of restriction levels — one per overload level. This rigid coupling of restriction and overload levels is not necessary (nor desirable) for good adaptive control in the authors' experience. This idea occurs, for example, in ITU-T Recommendation E.412 in connection with ISUP ACC, where two overload levels and two restriction levels are advocated.

7.8   Testing overload controls

Tests fall into two main types.

- Deterministic (calls are generated with constant inter-arrival times)

  This type of test is designed in such a way that the behaviour of the system is very precisely determined. It enables testing of whether:

  — the overload control logic is working correctly,

  — parameter values are being correctly assigned,

  — measurements (for statistics) are being correctly calculated.

- Realistic load and configuration

  These types of test are designed to determine the behaviour of the overload control when the test set-up is configured to mimic a more realistic network configuration, including the random nature of real traffic. Optimal or near-optimal parameter values would be used. It enables a determination of whether the observed behaviour sufficiently closely matches that determined by modelling (discrete event simulation), and hence whether the code implements the end-to-end overload control design faithfully.

In each of the above two cases, it is the end-to-end behaviour of the control that is being tested, and therefore such tests should cover both the overload detection and the restriction schemes.

The generic overload control requirements described in section 7.5 imply the following non-obvious testing and modelling facilities:

- call generators able to mimic the call state machine at source nodes, including the relevant parts of the overload control (if real switches are not available);

- equipment able to measure response times and calling rates (before and after restriction at source nodes, and admitted and rejected at the target node) on a second-by-second basis — required to observe the fast dynamics of the end-to-end control of powerful switches under overload;

- very high calling-rate generation (e.g. several thousand calls per second) to overload large capacity switches;

- a realistic model of the end-to-end overload control (usually a detailed simulation) — required to check that all parameter values cope well over the required range of overload scenarios;

- a comparison of test measurements with the simulation model's results — required if the full range of overload scenarios cannot be generated in a test facility.

7.9 Future overload controls

This section gives a brief discussion of possible, future, overload controls.

7.9.1 SIP and HTTP

SIP [26] and HTTP [27], from which it is derived, do not currently have any overload control methods that would meet the requirements presented here. However, they both have elements that could support the incorporation of overload controls designed according to section 7.6.

SIP is an application-layer control protocol that can establish, modify and terminate multimedia sessions (conferences) or Internet telephony calls. In particular, after sending an INVITE request message from a client to request participation in a session, there are various response messages returned which can be used in feedback control. Some of these relate to the progress of a request due to a server, and so can be used for server (nodal) load control. Others relate to a destination server status, e.g. busy or unavailable, and so can be used for overload focused on to a specific destination (using as identifiers 'To Header fields' or 'Request-URI'). Staged overload control can be applied because so-called intermediate nodes or proxy servers are used, for example, to route requests, enforce policies, and control firewalls.

HTTP has a similar structure, and therefore similar techniques can be used for restricting the rate of requests for a Web page. However, in the context of the WWW, caching [28] is an additional mechanism available to control overload to static pages. Rather than just restricting the rate of requests, copies of pages can be distributed away from the destination server hosting a page subject to overload, thereby reducing the path-length of requests and the load on servers hosting the original or cached copies.

7.9.2 Control of Release Avalanches

Repeat-attempt surges and mass-call releases need to be controlled in different ways. A stream of attempts can be throttled to a rate which matches the capacity of the network bottle-neck. Call releases, on the contrary, must not be discarded, because that would lead to over-charging some customers and leave network connections hanging. Moreover, if a call-release message is discarded, then the signalling system will (after a time-out period) retransmit it. If a high rate of release messages causes signalling delays longer than this time-out period, then a release avalanche can be triggered [29] in which retransmitted release messages lead to longer delays and more time outs.

A control could (in principle) be designed to adapt the time-out period so that the rate of such time outs is reduced to a low level. This would match the rate, at which release messages are sent, to the available capacity of the bottle-neck they traverse, which is the maximum sensible rate in the circumstances.

7.10   Summary

This chapter has presented evidence to show that overloads occur frequently in telephony networks, and can greatly exceed the capacity of the terminating lines (and of the network) to handle the surge of calls. The typical causes of overloads are media-stimulated events, emergencies and equipment failures. Their impact on the network is to reduce effective switch throughput (ultimately leading to switch failure) and to generate high levels of repeat attempts — most of which will fail to complete successfully, but nevertheless consume network resources, thereby reducing the capacity available to other, non-event, call streams. It is not economic to provide sufficient network or terminating line capacity to handle such events; consequently overload controls are necessary to ensure that switches are protected and ineffective traffic is minimised.

Overload controls need to be automatic, fast-acting and adaptive. As this chapter has shown, it is straightforward to define, in a generic way, how such controls should behave, without defining how they should be implemented to achieve that behaviour. Thus, for example, a key requirement is that the control should adaptively cause the admitted calling rate offered to an overloaded resource to converge close to the maximum rate it can handle, subject to keeping response times small enough to avoid customers abandoning calls in set-up.

It has been argued that such behavioural requirements should be a part of the specification of any control appearing in an international or UK telecommunications standard. Otherwise, there is no assurance that an implementation of a control will perform adequately. This view is beginning to be accepted in standards organisations, as recent experience at the ATM Forum and the ITU-T (and within the UK at PNO-ISC) has demonstrated.

The principles of good overload control design are now fairly well established. An overloaded system needs an internal overload control to reject an adaptively varied part of the stream of service requests offered to it. External controls are also required to adaptively regulate the demand offered to the overloaded system so that the total reject rate from the overloaded system is small but non-zero. The adaptation should be driven either by explicit rejection of service requests (accompanied by an indication of overload), or inferred from the expiry of suitable time outs.

These ideas have been successfully applied to design, test and deploy overload controls in BT's network.

The fitness of a control design can only be established by performance modelling, and the correctness of the implemented design can only be established by test measurements. This chapter has shown that non-trivial amounts of test equipment are required to adequately measure the end-to-end behaviour of a fast-acting control protecting a high capacity system from overload.

This chapter claims that all the basic control requirements and design principles discussed apply (maybe with small variations) across a wide range of network services and technologies, including, in particular, those based on Internet protocols.

The authors look forward to an increasingly widespread use of well-designed network overload controls in the future.

Appendix

7A   *Stability Analysis of a Class of Rate-Based Overload Controls*

The material in this appendix was originally written by one of the authors as a contribution to the BSI report [14]. The permission of the chairman of the PNO-ISC to reproduce it here is gratefully acknowledged.

7A.1   Steady State Analysis

Consider the class of overload controls defined in section 7.6.4. It is simple to analyse the steady state behaviour of all controls in this class. To do this, some notation is required. Let there be n source nodes, each being a point of restriction. Let γ_i denote the rate at which the restriction process at source i associated with the route to the overloaded node admits calls which are then offered to the overloaded node. Let:

$$\Gamma = \sum_{i=1}^{n} \gamma_i \qquad \qquad \text{...... (7A.1)}$$

denote the total calling rate offered to the overloaded node, and let $B(\Gamma)$ denote the probability that the overloaded node rejects a call from any source due to internal overload. The rate at which calls from source i are rejected by the overloaded node is given by:

$$\omega_i = \gamma_i B(\Gamma) \qquad \qquad \text{...... (7A.2)}$$

In equilibrium, when control is active at all sources, this reject rate equals the locally configured target reject rate at source node i, denoted by r_i:

$$\omega_i = r_i \qquad\qquad \text{...... (7A.3)}$$

From equation (7A.1), equation (7A.2) and equation (7A.3), it follows that:

$$R \equiv \sum_{i=1}^{n} r_i = \Gamma B(\Gamma) \qquad\qquad \text{...... (7A.4)}$$

where R denotes the sum of the individual source node target reject rates.

If the overloaded node's call rejection probability $B(\Gamma)$ is a continuous and increasing function of the total calling rate Γ offered to the target, there will always be a unique value of Γ (denoted by Γ_s) where R/Γ and $B(\Gamma)$ intersect and which hence satisfies equation (7A.4), as illustrated in Fig 7A.1.

In Fig 7A.1, L_M is the offered load at which the overloaded node's effective throughput is maximised, and L_C is the offered load beyond which correct call handling is not guaranteed — see section 7.3.2. It is clear from Fig 7A.1 that if R is a small fraction of L_M, then the equilibrium value of Γ will be close to L_M, since reducing the value of R lowers the curve of R/Γ. Thus it may be concluded that the steady state behaviour of this class of overload control (if it is achieved) gets close to maximising the effective throughput of an overloaded node, whatever the overloaded node's capacity L_M may be, and however many source nodes (restriction points) there are.

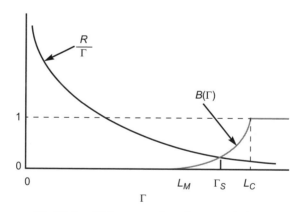

Fig 7A.1   Uniqueness of steady state solution.

In addition to this, we show below that source i gets a share of the overloaded node's effective capacity which is proportional to source i's target reject rate. To see this, observe first that the share of the overloaded node's capacity that source i gets is given by:

$$\upsilon_i = \gamma_i(1 - B(\Gamma_s)) \qquad\qquad \text{...... (7A.5)}$$

effective calls/sec.

From equation (7A.2) and equation (7A.3) this equals:

$$\upsilon_i = r_i \cdot \frac{1 - B(\Gamma_s)}{B(\Gamma_s)} \qquad \text{...... (7A.6)}$$

Summing this equation over i gives the total effective throughput of the overloaded node:

$$\upsilon = R \cdot \frac{1 - B(\Gamma_s)}{B(\Gamma_s)} \qquad \text{...... (7A.7)}$$

Equation (7A.6) and equation (7A.7) show that:

$$\upsilon_i = \frac{r_i}{R} \cdot \upsilon \qquad \text{...... (7A.8)}$$

That is, source i gets a share of the overloaded node's effective capacity which is proportional to the target reject rate at source i.

The analysis in this sub-section has tacitly assumed that all call processing at the overloaded node occurs at call set-up. It would be more realistic if the model allowed the processing to be distributed over the call's lifetime, e.g. with some processing at call set-up and more processing later at call clear-down. It can be shown that very similar results apply in that more realistic case.

7A.2   Convergence to Steady State

Suppose that the overloaded node is subject to high calling rates from each of n source nodes (each acting as a restriction point) from some point in time taken to be $t = 0$. We assume that the transient behaviour of the set of overload controls is adequately described by the following set of ordinary differential equations (ODEs):

$$\frac{d}{dt}\gamma_i(t) = f_i(r_i - \omega_i(t)) \qquad \text{...... (7A.9)}$$

This says that the rate of change of $\gamma_i(t)$ at time t is some function of the difference between the detector reference reject rate r_i and the rate at which source i's admitted calls are rejected by the overloaded node. This rate, which is denoted by $\omega_i(t)$, is given by the equation:

$$\omega_i(t) = \gamma_i(t)B(\Gamma(t))$$

where $\Gamma(t)$ denotes the total rate at which call set-up requests are admitted by the restriction points and offered to the overloaded node:

$$\Gamma(t) = \sum_{i=1}^{n} \gamma_i(t)$$

The function f_i approximately describes the combined effect of the three components (U, D and R — see section 7.6.3.3) of the overload control instance at source node i. It may vary from source node to source node (e.g. because they come from different suppliers). The only conditions placed upon it at this point in the analysis are that it is continuous and takes the value 0 only when $r_i = \omega_i(t)$. This ensures that in equilibrium (i.e. when all derivatives $d\gamma_i(t)/dt = 0$) we must have:

$$r_i = \omega_i(t) \qquad\qquad ...... (7A.10)$$

for all $i = 1, ..., n$.

The set of ODEs given by equation (7A.9) is almost a special case of the ODEs considered in Kelly et al [7] as part of their stability analysis of sets of Internet overload controls. The only difference between the equation (7A.9) set and those considered in Kelly et al [7] is that they consider the case where all the functions $f_i(x) = \kappa x$ for a positive constant κ. It turns out that their stability analysis [7] still carries through successfully provided only that a mild additional constraint is placed upon the functions f_i. That constraint is that $f_i(x)$ is positive when x is positive and negative when x is negative.

This is intuitively reasonable, since it just says that the control instance at source node i increases its admitted rate $\gamma_i(t)$ at time t if $r_i > \omega_i(t)$ and reduces it if $r_i < \omega_i(t)$.

Translated to the set of feedback controls given by equation (7A.9), the method used in Kelly et al [7] to establish stability basically first proves that the function:

$$V(\gamma_1, ..., \gamma_n) = \sum_{i=1}^{n} r_i \log\gamma_i - \int_0^{\sum_{i=1}^{n}\gamma_i} B(y)dy \qquad ...... (7A.11)$$

is strictly concave [32] on the set where all $\gamma_i > 0$ and hence has just a single global maximum, attained at the unique point where its gradient vanishes:

$$\frac{\partial V}{\partial \gamma_i} = \frac{r_i}{\gamma_i} - B\left(\sum_{j=1}^{n} \gamma_j\right) = 0 \quad i = 1, ..., n \qquad ...... (7A.12)$$

Then, it is shown that the derivative of V along a solution trajectory:

$$\frac{dV}{dt} = \sum_{i=1}^{n} \frac{\partial V}{\partial \gamma_i} \frac{d\gamma_i(t)}{dt} = \sum_{i=1}^{n} \left(\frac{r_i}{\gamma_i t} - B\left(\sum_{i=1}^{n} \gamma_j t\right)\right) \cdot f_i\left(r_i - \gamma_i(t) B\left(\sum_{j=1}^{n} \gamma_i(t)\right)\right)$$

$$...... (7A.13)$$

is positive (except where V attains its maximum where dV/dt is zero) provided that $f_i(x)$ is positive when x is positive and negative when x is negative. So, along any solution trajectory, the feedback controls jointly maximise V, and all trajectories must converge to the unique point which maximises $V(\gamma_1, ... \gamma_n)$ and which is characterised by equation (7A.10).

We may therefore conclude that, provided $f_i(x)$ is positive when $x > 0$, zero at $x = 0$, and negative when $x < 0$, the overload controls converge globally to the unique steady state discussed in section 7A.1. This is very important, because it says that different restriction points may implement the controls in different ways (as characterised by their different functions f_i), but convergence to the unique steady state is nevertheless guaranteed.

Now, of course, real overload controls cannot be exactly described in this way, because the analysis has not taken into account the effects of, for example, signalling and processing delays, and the stochastic nature of demand.

Consequently the constraints on the functions f_i must be regarded as sufficient for convergence of the set of overload controls given the use of ODEs in this section is a valid approximation, rather than generally sufficient. In practice, more detailed modelling would be required. This sort of initial stability analysis is a useful guide to what results could be expected from more detailed simulations that take delays and stochastic effects into account.

References

1   European Patent EP 0 729 682: '*A method of controlling overloads in a telecommunications network*'.

2   Network Interoperability Consultative Committee: '*PNO-ISC Specification Number 007: ISDN User Part (ISUP)*', Issue 3.0 (July 2001).

3   ATM Forum: '*Signalling Congestion Control*', version 1.0, ATMF ref af-cs-0181.000 (2001).

4   Williams, P. M.: '*A novel automatic call restriction scheme for control of focused overloads*', 11th UK Teletraffic Symposium, Cambridge, IEE (1994).

5   Rumsewicz, M. P.: '*Critical congestion control issues in the evolution of common control signalling networks*', Proceedings of the 14th International Teletraffic Congress, Antibes Juan-les-Pins, France, Elsevier (June 1994).

6   Wischik, D.: '*How to mark fairly*', Final Draft (Jan 2001) — http://www.statslab.cam.ac.uk/~djw1005/Stats/Research/marking.html

7   Kelly, F. P., Maulloo, A. K. and Tan, D. K. H.: '*Rate control for communications networks: shadow prices, proportional fairness and stability*', Journal of the Operational Research Society, **49**, pp 237-252 (1998).

8   Finkel, R. A.: '*An operating system's vade mecum*', Prentice-Hall (1988).

9   Korner, U.: '*Overload control of SPC systems*', 13th International Teletraffic Congress, Copenhagen, North-Holland (1991).

10 Vahalia, U.: '*Unix internals*', Prentice-Hall (1996).

11 Pre-Published ITU-T Recommendation Q.714 (05/2001): '*Specification of Signalling System No. 7 — Signalling Connection Control Part*', (2001).

12 Miller, R. K. and Michel, A. N.: '*Ordinary differential equations*', Academic Press (1982).

13 Kalman, R. E. and Bertram, J. E.: '*Control Systems Analysis and Design via the second method of Lyapunov I: Continuous-time systems*', Journal of Basic Engineering, pp 371-393 (1960).

14 British Standards Institute, PD 6673:2001 (PNO-ISC/INFO/015): '*Full report on ISUP overload controls*', (2001).

15 Whitehead, M. J.: '*Adaptive ISUP overload control*', ITU-T, Study Group 2, Copenhagen, Temporary Document CPH19.

16 US Patent US 5450483: '*A method of controlling overloads in a telecommunications network*'.

17 Network Interoperability Consultative Committee: '*PNO-ISC specification number 010 bearer independent call control protocol (BICC)*'.

18 ITU-T Study Group 2 — http://www.itu.int/ITU-T/studygroups/com02/index.html

19 ITU-T Study Group 16 — http://www.itu.int/ITU-T/studygroups/com16/index.html

20 ITU-T Recommendation E.744 (10/96): '*Traffic and congestion control requirements for SS No. 7 and IN-structured networks*', (1996).

21 ITU-T Recommendation Q.1902.4 (07/01): '*Bearer independent call control protocol, basic call procedures*', (2001).

22 ITU-T, H.248 Annex M.2: '*Media gateway resource congestion handling package*', (2001).

23 British Standards Institute, PD6645:2002: '*PNO-ISC specification number 006 — interconnect user part (IUP)*', (2002).

24 ITU-T, Q.764 (12/99): '*Signalling System No 7 — ISDN user part signalling procedures*', (1999).

25 ITU-T, Q.724 (11/88): '*Specification of Signalling System No 7 — telephone user part*', (1988).

8

REALISING EFFECTIVE INTELLIGENT NETWORK OVERLOAD CONTROLS

P M Williams and M J Whitehead

8.1 Introduction

The original idea of an 'intelligent network' (IN) was to be able to define a system capability that would support the rapid building and deployment of a large range of new services into a telephony network. Services include those with advanced call distribution features, such as call queueing. The ITU specifies a series of recommendations that define such a capability, incorporating the intelligent network application protocol (INAP) [1]. Abernethy and Munday [2] provide a useful overview of IN standards and services.

8.1.1 Overload Causes, and What is Special about an Intelligent Network

Although overloads can occur simply because insufficient capacity has been provided for the demand expected, it is more common for them to be caused by some quickly arising event, which may be unanticipated. These include network or system failures, tariff changes, and network system processes that have been scheduled to occur synchronously. They also include the types of service often deployed on an IN, such as media stimulated events or natural emergencies (e.g. bad weather), when the traffic can be very volatile. To make matters worse, traffic is usually magnified by calling customer or system repeat attempt behaviour. Such overload causes are discussed further in section 7.2 of Chapter 7. Any system needs effective overload controls in order to avoid excessive response times and reduced throughput, or even failure (see Chapter 7 for more details). In the case of an IN, there are architectural reasons why it may be especially susceptible to overload, and these are explained below.

The essential components of the ITU IN functional model are shown in Fig 8.1. The service control function (SCF) contains call control functions composed of the fundamental building blocks of services. These interact with service logic and data, and interface to other functions including the service switching function (SSF) and specialised resource function (SRF). The SSF extends and modifies call control functions to allow recognition of IN service control 'triggers' in order to query the SCF, and manages signalling between the call control and SCF. The signalling message resulting from a triggered 'detection point' is an initial detection point (IDP), sent from the SSF to the SCF.

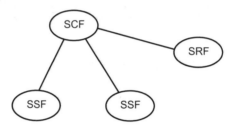

Fig 8.1   Some components of the IN functional model.

The SRF provides the resources required for the execution of IN-provided services, e.g. digit receivers and announcements.

The node which hosts the SCF is usually referred to as a service control point (SCP), and that which hosts the SSF, i.e. a switching system, as a service switching point (SSP).

While the ITU standards specify an IN in a distributed manner, independent of physical realisation, a typical IN architecture is centralised in the following sense: there are only a small number of SCF instances, and many instances (possibly hundreds) of SSFs per SCF. Such high connectivity, in combination with the usual capacity of each SSP, means that the total demand that can be generated by the SSPs can easily be much greater than an SCP's capacity. Furthermore, the total SCP capacity is usually much greater than that of (say) a destination switching system to which it instructs connections to be made. Thus if an overload is focused in nature, i.e. the traffic is concentrated on to one or a few destination systems, then there are two possible consequences:

- a destination or the intermediate network may not have adequate overload controls to limit the load processed, resulting in degraded performance (excessive response time or reduced throughput);

- even if controls are adequate, the bulk of the calls processed by the SCFs may be ineffective because the destination resources (lines) are busy; the SCFs would then be processing a large amount of ineffective workload, which may be causing other calls to be rejected that would otherwise have had a good chance of completing — the effective throughput of the SCFs would therefore decrease.

8.1.2   Example of Node Failure

Suppose an IN is composed of several SCP nodes. A typical scenario, which can give rise to an overload, occurs when one of the nodes fails and its traffic has to be re-routed to working nodes. To be specific, assume that there are three nodes, each serving many SSPs (see Fig 8.2). If node 1 fails, then the SSPs re-route their traffic (fresh calls) to the two remaining nodes, half of them to node 2 and half to node 3. If the load before failure was approximately evenly split over the three nodes, then after failure the load on nodes 2 and 3 increases by a half, which may cause overload if it occurs at a busy time of day.

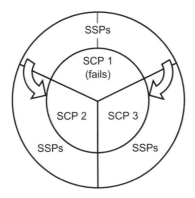

Fig 8.2   Re-routing of traffic from SSPs to SCPs on node failure.

This scenario will be considered later in terms of setting control parameters and analysing performance (see section 8.6.2.2).

8.2   IN Standards

8.2.1   What the Standards Provide

In developing the recommendations for IN standards, the ITU recognised that overload of the SCF could easily occur. They therefore developed recommendations to support the use of overload control on the SSF-SCF interface. This is the 'call gap capability' [1], so named because it specifically requires the call gapping restriction method (described below). There is also a so-called 'service independent building block' (SIB) that is provided by the standard, called the 'limit SIB', to enable call counts for televoting to be made at SSFs, and transferred in batches to the SCF. This limits SCF queries. However, this only provides a load control mechanism for a specific service type (vote counting), and in reality many service providers wish to perform their own counting and therefore do not wish to use this capability.

The call gapping restriction method (at the SSF) works as follows. Each time a call (which has triggered an IN query as an IDP at the SSF) is admitted, a timer is started of length equal to the 'gap interval'. While this is active, all calls (that match the criteria specific to the control instance) are rejected. When it expires, the next arriving call is admitted and the timer restarted. There is also a longer 'duration' timer. This determines how long control is active. This is restarted each time a call gap message is received. Control terminates only when this expires, or a message is received with the duration set to 0. In section 8.6.1.2 we will provide an analysis of call gapping.

Gapping control applies to calls which meet 'gap criteria' specified by the SCP, which include a service identity known as the 'service key' (SK), or leading digit substrings of the 'called party number' (CdPN) or 'calling party number' (CgPN). Multiple controls to different criteria may be active simultaneously at an SSF. These gap criteria, together with the control parameters 'gap interval' and 'duration' and other information elements, are carried by the call gap capability information flow from the SCF to the SSF.

8.2.2   What is Not Provided by the Standards

The ITU standards provide a mechanism for carrying control parameter values specific to the gapping algorithm from the SCF to SSFs. However, this is only part of providing an overload control solution. It is only directly concerned with controlling load inbound to the platform, and does not concern SCP internal overload control (see section 7.6.1 in Chapter 7), or outbound control (see section 8.5.3.1) at all.

Even with regard to controlling inbound load, it still only provides a small part of the solution. Essential parts of a control are the functions that detect and monitor overload status and update the control parameter values. These are not addressed at all by the call gap capability standard.

It is therefore necessary for network operators or suppliers to complete the solution.

8.3   BT's IN Implementations

BT has a range of services that are provided by IN platforms.

Horizon is a high-capacity platform. It supports a range of basic and advanced services offering features such as proportional call distribution over multiple sites, geographic and time-of-day routing, alternative routing on destination busy, call queueing, announcements, etc. Customers providing the services can have near real-time statistics concerning call outcome. It also integrates a universal card services platform.

The 'intelligent call manager' (ICM) is a system supporting advanced service features such as call queueing, with the provision of detailed call progression statistics and answering agent status. These platforms benefit from having many of the features that are described in this chapter.

8.4   Requirements for IN Overload Controls

8.4.1   General Requirements for the Effect of Controls

The generic overload control requirements, discussed in detail in section 7.5 of Chapter 7, may be summarised as follows:

* maximise effective throughput (in this chapter, an effective call is one that results in the seizure of a terminating line) — if there is sufficient demand, then such terminations should be kept highly utilised in order to maximise revenue, satisfy the service provider who hosts the lines, and increase the chances of a caller making a successful connection;

* bound response times — excessive response times lead to abandoned calls, and therefore wasted processing (details can be found in Chapter 7);

* limit ineffective calls (in this chapter ineffective calls will be taken to include calls that fail due to busy network terminations, voice circuit route exhaustion, signalling network congestion, and rejection by system admission control; calls that result in alerting a called party, but are not answered, are not included) — during overload, ineffective calls waste network capacity and so must be limited.

In addition to these basic requirements, it may be appropriate for the control to behave in specific ways relating to:

* different call priorities;

* fairness (in some sense) among controlled sources.

8.4.2   Inbound and Outbound Control

An overloaded SCP should be able to activate call restriction at SSPs. This is called inbound control since it throttles calls inbound to the SCP.

An SCP can heavily overload network resources (e.g. switches and answer centres), and so needs a separate control to protect them. This is called outbound control since it throttles calls outbound from the SCP to specific identified destinations.

In order to limit ineffective traffic in the SCP itself, which could result from overloaded network destinations, it is also necessary to apply inbound gapping at SSPs on IN triggered calls to specific called numbers.

8.4.3   Intelligent Network Application Protocol Standards

A practical SCP overload control must build on existing standards if it is to be readily adopted by suppliers. In particular, it must use the 'IN call gap facility' of INAP (see section 8.2.1). For control of traffic inbound to the SCP, this implies that call restriction external to the SCP must take place at source SSPs, using the call gap algorithm. Detection of SCP overload and updating of the IN gap interval must take place at the SCP itself.

8.4.4   Measurements

There are several reasons why it is necessary for an overload control to provide measurements that indicate what it is doing. These include:

- 'near real-time' network management — it is useful for network managers to have not only a list of any call stream identities that are under control, as this may facilitate diagnosis of other network events or conditions that are observed, but also the ability to provide, on demand, more detailed measurements, such as call counts and their outcomes, and control parameter values;

- optimisation of control parameter values based on observed control behaviour;

- diagnosis of faulty operation of the control while in service.

We will show how to derive some control parameters based on a theoretical analysis, but a general consideration of control measurements is beyond the scope of this chapter.

8.4.5   Testing

It is important that, once realised, automatic controls of the type being recommended here must be properly tested, to show that they conform to the design, and to verify that the design is effective in a realistic test network configuration. A more detailed consideration of testing such controls is presented in section 7.4.8 of Chapter 7.

8.5   IN Overload Control Design

From the overload control requirements presented above in section 8.4 (and also in section 7.4.5 of Chapter 7), it is possible to derive good design principles which ensure that the requirements are met. These have been used as the basis for the designs presented in this section.

8.5.1   Static or Adaptive Control?

Controlling overloads by adapting the level of call restriction allows the rate at which calls or connections are admitted to be just sufficient so as to maximise the occupancy of resources (with effective load). Static load controls can lead to over- or under-control. They are particularly unsuitable in this respect for inbound control to an SCP, where there are many source SSPs. In this case, for a given gap interval, the admitted rate can vary enormously according to the distribution of traffic over the SSPs (see section 8.6.1.2). They can be used with more success for outbound control (see section 8.5.3.1), where there are a smaller number of points of restriction on the SCP nodes, assuming that the load is distributed fairly uniformly over these. The biggest problem in this case is administrative, because the static rate has to be set according to an estimate of the destination mean holding time and the number of terminating lines, both of which can vary over time. The design that will be presented here is based upon the use of adaptive controls throughout. However, deploying static outbound rate controls and then progressing to the adaptive version would provide an attractive approach to phased introduction of the fully integrated and adaptive controls.

8.5.2   Basis of the Design

The IN control scheme described here is based upon an adaptive overload control method designed and patented by BT [3] which will automatically control focused overloads in any telephony network, including IN. It has variously been referred to as 'automatic call restriction' (ACR) or 'dynamic call restriction' (DCR), and satisfies the requirements for 'automatic destination control' (ADC), specified in ITU standard E.412 [4].

It uses the principle that the rate of call failure (downstream) is a very effective way of detecting overload (see section 7.6.3.2 of Chapter 7 and Langlois and Régnier [5]). The control enforces an upper limit to this rate by activation and updating of call restriction at appropriate points (upstream). Under steady load the controlled failure rate will converge to this by adaptation of control parameters, assuming a sufficiently high level of offered traffic (otherwise the control will turn off).

It will be assumed that network systems, including the SCP in the IN context, have an admission control function (see section 7.6.1 of Chapter 7) (denoted hereafter by A). This limits the rate at which new calls are accepted, in order that the system maintains acceptable response times, e.g. by monitoring response times or occupancy. Although it is necessary for the control, the design of this particular function is not within the scope of this chapter. Ineffective calls monitored by the control include those rejected by A (for inbound control), and those which fail while trying to be connected to a network destination (for outbound control).

In addition, it is necessary to distinguish between different traffic streams, because an overload may be predominantly caused by calls to just one or a few numbers, which therefore have a small chance of being successful at their destinations. Other call streams (which share the same network resources) may have a good chance of being successful at their destinations. These should only be controlled once the offending call stream is appropriately controlled. The control that will be presented here can be applied with any specified criteria to distinguish call streams.

In telephony networks, the most useful call stream identities to use are node identity, for controlling overloads to a network node, and the called number, for controlling overloads focused on to specific destination resources. For control inbound to an SCP it is the SK, and CdPN or CgPN, that can be used, because of what is provided by the IN call gap capability (see section 8.2.1). For control of demand outbound from an SCP, the destination routing address (DRA) is used.

The design and performance of ACR, as applied to telephony networks in general, has previously been reported [6], but it is useful to summarise here the functional components of the scheme before describing their use in an IN context. These are common to any feedback type of call restriction control that uses call failure rate as a control variable (see section 7.6.3.3 of Chapter 7). Figure 8.3 shows how the functions of the control, denoted D, U and R, and described below, may be configured in coupled control loops with admission and routing control functions A. Later, Figs 8.4 and 8.5 will show how these functions may be mapped on to IN components.

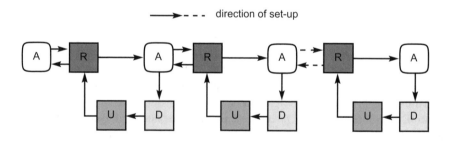

direction of set-up

A - admission and routing control
D - detection and monitoring
U - updating restriction parameters
R - restriction

Fig 8.3   ACR component functions, D, U and R, configured in coupled and staged control loops.

8.5.2.1 Overload Detection and Monitoring Process (D)

This function counts calls that have failed downstream as a means to detect and monitor overload. It consists of a set of leaky bucket monitors that can be dynamically assigned to call stream identities (these were discussed above in section 8.2.1 for the IN context). The fill of each monitor bucket leaks at a constant rate equal to the target rate of failed call/connection. Whenever an indication of a call failure is received, the set of monitors is searched to find one that is already assigned to a stream identity that matches that of the failed call. If a match is found, then the fill of the monitor is increased, otherwise an unassigned monitor is associated with the failed call stream identity and its fill initialised.

The status of each monitor is determined by the changes to its fill relative to a set of thresholds, so that it may be in a state of high or low occupancy. If a change from a low to a high occupancy occurs, it is an indication to start restriction or increase the level of restriction against the stream identity to which the monitor is assigned. Conversely, if a change from a high to a low state occurs, then it is an indication to decrease restriction.

8.5.2.2 Restriction Level Updating (U)

Increasing the level of restriction is taken to mean that a call is more likely to be rejected (for a constant offered load), and vice versa for decreasing the level. Adaptation of the restriction level is started when the first indication to increase the level of restriction is received as a result of a monitor onset threshold being crossed. Thereafter it is increased when further such indications are received, subject to a specified maximum level. Similarly, whenever decrease indications are received, the level of restriction is decreased until a minimum is reached.

At this point, the control attempts to turn off restriction, but the updating function remains active until a long timer expires (the reasons for this are explained in section 8.5.4.1).

8.5.2.3 Restriction (R)

The output from the restriction updating function is updated restriction control parameter values, to be distributed to the points where call restriction is actually applied. Several different restriction algorithms can be used. Inbound control must use call gapping (see section 8.2.1), because it depends upon the IN call gap capability information flow. For outbound control restriction (see section 8.5.3.1) using a leaky bucket (see section 7.6.3.6 of Chapter 7), with the leak interval as the adapted parameter, is recommended.

8.5.3 Inbound, Outbound and Staged Control

Figures 8.4 and 8.5 illustrate the configuration of the ACR control on an SCP node. The 'service logic' entity is a representation of where the plans for the various services are executed. We will assume that an effective overload control internal to the SCP is used, represented by the admission control function A.

Figure 8.4 shows the configuration of the control when an 'inbound control function' (ICF) alone is used. This comprises an instance of the ACR control functions D and U. D uses indications of call rejections from A, and U controls updates to the gap interval. Gapping is applied at R in the SSF. Some of the issues concerning call gap distribution are discussed in section 8.5.4.2.

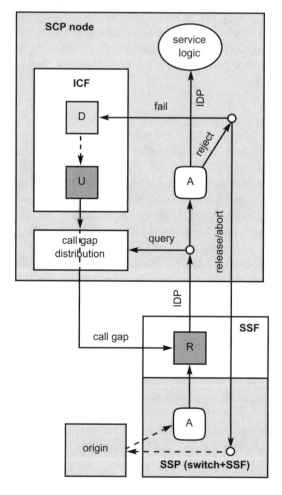

Fig 8.4   Functions and information flows for inbound overload control.

This adequately protects the SCP but does not necessarily protect network destinations. It is therefore also desirable to have an 'outbound control function' (OCF) to control demand sent to the network, as illustrated in Fig 8.5. This comprises all the control functions *D*, *U*, and *R*, which may be distributed across several SCP nodes.

Inbound and outbound controls are therefore implemented as two distinct control loops, i.e. two instances of ACR, the ICF and the OCF respectively. They may be coupled by the reason that calls rejected by the OCF (because all possible connection attempts have been exhausted by service logic) are indicated to the detection and monitoring function of the ICF (see Fig 8.5 and section 8.5.3.2). Staging the controls in this way has a number of advantages that are discussed further in section 7.6.3.4 of Chapter 7. In addition, it should be noted that:

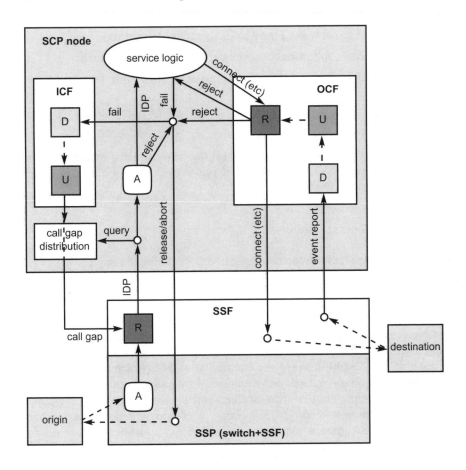

Fig 8.5   Functions and information flows for integrated inbound and outbound overload control.

- call gapping will be invoked less often at the SSPs (only for the more severe events) than would be the case if it were activated directly by network destination congestion, because there is a limited rate at which calls can be ineffective due to OCF restriction in the SCP itself — this will limit signalling traffic to SSPs carrying call gap messages;

- translation from CdPN (and possibly CgPN) to a 'destination routing address' takes place in the SCP, as determined by service logic.

There are several references in the following sections to calls being rejected as a result of restriction control, activated due to a destination with very high occupancy. What tone or announcement should be played to a caller? The following arguments are for playing a network congestion tone or announcement:

- it is the network that has rejected the call in order to avoid network congestion;

- the call has not actually met a busy condition at the destination.

On the other hand, the following are arguments for playing a busy tone:

- the call would have had a very high probability of meeting busy had it been admitted;

- true network congestion may not have actually occurred, in the sense that network resources are in overload.

Another factor is the type of events from the network being monitored by the control (see section 8.5.3.1). Therefore, it is not possible to be dogmatic about which tones should be applied.

8.5.3.1   Outbound Control

Use of Event Reports for Adaptive Outbound Control Feedback

Adaptation of outbound control can be achieved by using INAP messages to monitor when connection attempts have failed due to network congestion or destination busy.

Connection attempts to network destinations are initiated by service logic requesting that the SCF sends to the SSF the INAP operations 'connect' (or 'connect to resource'), or 'establish temporary connection' (ETC) (which is used to connect to the SRF). Detecting overload is accomplished by instructing the SSF to monitor for events that imply busy destination (lines) or intermediate network congestion. To do this a 'request report BCSM event' operation is added to the connect operation, which carries a request to monitor for destination busy or network congestion. These conditions are defined by two event types, 'called party busy' or 'route select failure', respectively. If the ETC operation fails then a message indicating this will be provided by the SSF.

It is possible that the network already has reliable and effective switch overload controls that discriminate call streams with a small chance of leading to effective calls (on the basis of network address). If this were so, then it would only be necessary to monitor for network congestion.

However, this is often not the case, which means that monitoring on destination busy is necessary. For random connection attempts, control target (busy) rates of between 2 and 10 calls/sec will give destination line occupancies of between 95% and 99%, for any size of line group and mean holding time greater than 10 sec (see section 7.6.3.2 in Chapter 7). These are target rates for all SCP nodes making connection attempts, and therefore with several instances of OCFs the individual monitor leak rates would have to be assigned accordingly with knowledge of how load is distributed over SCPs.

It is possible that some (destination) customer switching systems route calls with the same destination address to several line groups, but they do not balance the load very evenly. This might happen when static proportional distribution of traffic is used with inaccurate proportions and without alternative routing. In extreme cases, when line groups are large and with short call-holding times, this could lead to activation of control when in fact one of the line groups does not have a high occupancy. There are a number of ways to circumvent this potential problem. One of the simplest is to provide a default monitor connect failure rate for all destinations, but override this with a suitably higher rate for destinations known to have the above characteristics.

Event reports can be requested in two modes — 'notify and continue', or 'interrupt'. For notify and continue it is an instruction to the SSF that when an event of the specified type has been detected it should be reported to the SCF but then call processing should continue as normal.

For interrupt, when an event is detected by the SSF, processing is interrupted, and, after sending the SCF an event report, the SSF awaits further instructions from the SCF.

If the attempt is admitted by the OCF, and relevant event report types are not already present, then they must be added, with the mode notify and continue. Received event reports of either type are counted with a leaky bucket monitor (in the control function D), dynamically assigned to the destination routing address.

For most IN service mixes, the signalling message load on outbound signalling links from the SCF is greater than that in the inbound direction (typically more than twice as much, although somewhat less for very basic services). Because of this, there may be concern that the outbound message load should not be increased by the addition of a request to report an event for every message. However, the following should be noted:

- as complex services, or those where calls require statistics on the outcome of the call, will often send a request report in any case, the additional load for mixes with a high proportion of such calls is less;

- it is better that a known bound applies to the outgoing signalling load, which will be the case when outbound control is used, rather than an unbounded (and largely ineffective) load that can result without such control.

In any case, a number of simple techniques can be used to reduce this load in the design of the control, including:

- sample the requests, i.e. 1 in n, for services that have not already requested a report — in this case the load is always present but it is a lower proportion of the total;

- monitor the rate at which streams of connection request are sent to each destination address and only add request report when this rate becomes greater than the target control failure rate, thus reducing the signalling load incurred by destinations with a capacity (rate) to terminate calls lower than the monitor leak rate — this requires that all outgoing connection attempts must be counted and monitored against the network address rather than only those for which an event report is received.

Independence from Service Features

It is important that the OCF does not cause any undesirable interactions with the execution of service logic, which may include various alternative routing or load-sharing algorithms. The way that this can be achieved is explained below.

Where several destinations may be attempted (or re-attempted), service logic may monitor for the events described in the previous section that indicate network congestion or destination busy (or both). Which are monitored will depend on exactly how the services have been defined, and therefore the distinction between the two event types will no longer be made explicit in this section.

Outbound control functions can either be incorporated into service logic as additional service features, or maintained largely independent from services as a separate network management function which is called by each service, and where monitoring and control is performed. The advantage with the latter approach is that the overload control function can be separately designed, developed, and tested, and there is a clear division between service functions and network management functions. This can be done provided that the control function is designed so that connection attempts rejected by it do not interfere with any alternate routing or load sharing/balancing used by the services. This design will therefore be assumed in the following.

When the OCF receives a connection attempt from service logic, it will first encounter the restriction (admission control) function. If the attempt is refused by the OCF, then it should examine whether any event report of appropriate type has been requested, and, if so, an appropriate event report returned. This allows service logic to try some other action, such as an alternative destination. If no report is requested, or the mode is of type notify and continue (not interrupted), then the destination must be the last alternative, and service logic assumes that the call will be cleared down by the network (SSF). Since the connection attempt has been refused by the control, it is necessary to clear down the call, e.g. by sending release call, or by sending connect to an announcement facility.

If the attempt is admitted, then a request report is added as described above if it is not already present.

Consider how best to configure overload control with regard to services using alternative routing to network destinations when the destination is busy or the network congested. Suppose that a particular service re-routes calls to up to n points $A_1, A_2, ..., A_n$ in the network in turn if destination busy or network congestion is met (i.e. if access fails to access A_i, then A_{i+1} is attempted for $i = 1, ..., n-1$, where n has to be sufficiently small to avoid excessive post-dial delays).

Suppose first that the rate of call attempts to the service is controlled (i.e. effectively the rate at which IDPs are accepted) according to the rate at which connections fail after trying the last destination A_n. Then the traffic sent to A_i would include all that which is successful at $A_{i+1}, ..., A_n$ plus the controlled overflow rate on A_n. Therefore the rate sent to the earlier destinations tried would be much greater than the later ones, and have a larger proportion of calls which are ineffective. This is clearly undesirable from an overload control viewpoint.

A better solution is to provide call detection and restriction for the stream of connection attempts to each destination. Then the rate sent to A_i (after restriction) is just the sum of the rate of calls that successfully terminate on A_i and the rate that find A_i busy. The latter rate is limited by the control.

8.5.3.2 *Configuration of Overload Detection and Monitoring Functions*

Inbound Control Only

If the OCF is not present or not linked to the ICF by rejected connection attempt indications being sent to the ICF, then there are two fundamentally different approaches to configuring the monitor (leaky) buckets of the ICF.

Consider just having one nodal monitor for the whole SCP site, using an SK (see section 8.2.1) as node identifier. Depending upon how services are deployed across different SCPs, several SKs, and therefore several monitors, may be required (because of the limitations of the INAP call gap capability), but the analysis of this is similar.

A nodal monitor provides 'number-independent' monitoring, i.e. independent of CdPN or CgPN. All indications of calls rejected by SCP admission control are passed to the function D of the ICF for the fill of the nodal monitor to be increased, and gapping activated and updated accordingly. It is easy to show (see section 8.6.1.1) that traffic streams to different numbers see approximately the same probability of being rejected by call gapping or SCP admission control. Each stream therefore gets a slice of the node's capacity that is proportional to its size (rate). This can be viewed as fair, but it does have the drawback that if a stream grows very large it can dominate the node's capacity. This might be appropriate if the capacity to terminate calls in the network is very large, but in most instances this is unlikely, in which case the level of ineffective traffic may be large.

A way around this is to have sets of monitors, made up of a single nodal monitor with a large leak rate, R, and a set of n monitors that are dynamically assigned for number-specific control. Each of these has a leak rate, r, satisfying $nr < R$. It can then be shown (see section 8.6.1.1) that the available site capacity is divided up equally between each stream under control, so that no one stream can dominate. However, there still may be a large number of calls which lead to ineffective connection attempts in the network, and the way to overcome this, with the current control, is to link outbound to inbound control.

Inbound and Outbound Control

The aim of adaptive outbound control is to limit the rate of connection to the network to be just sufficient to maximise destination occupancy, subject to the other constraints in section 8.4.1. The previous section showed that the effect of inbound control is to allocate SCP capacity either in proportion to different stream size, or in proportion to the (specific number) monitor leak rates. There will therefore be a mismatch between outbound and inbound, in general. This can lead to some streams being overrestricted, and therefore their destinations being under-utilised. This is very undesirable. However, outbound control can be coupled to inbound control, by using the rejection of calls by the OCF as input to the overload monitoring of the ICF. This overcomes these problems, but there are some factors to take into account, which are discussed below.

Many factors can determine which destination routing address is assigned to a connection attempt by service logic, but the only ones that can be used by the ICF are SK, or leading digit substrings of CdPN, and CgPN (see section 8.2.1). Here, we will refer to the combination of these that the service uses to determine the routing address as the 'call translation context' (CTC). Typically it is just an SK and CdPN, but a substring of the CgPN may be used for geographical-based routing.

Outbound control is coupled to inbound control by passing the CTC to the ICF with an indication of connect rejection. This must only be done for OCF rejects that occur after service logic has completed call processing (dropped-call context), i.e. it represents complete call failure (see section 8.5.3.1). It is only passed to a complete number monitor, because it is necessary for the ICF to control just this stream. This is important because otherwise, if there is a generally high level of ineffective connection attempts in the network but spread over many destinations, inbound control could be activated when in fact none is necessary. This would lead to under-utilisation of network destinations.

What is also different from uncoupled outbound control is that call attempts rejected by SCP admission control are passed to the ICF without an indication of CTC, or equivalently these rejects are only registered by the nodal monitor.

A summary of three ways in which ICF monitors can be configured is given in Table 8.1. A ✓ means that monitors are provided to register call rejections from the specified function, and **X** means that they are not.

Table 8.1   How call rejects are monitored by the ICF for different control configurations.

Monitor type	Call rejection function	Inbound nodal only (1 monitor)[*]	Inbound (nodal and numbers)[*]	Inbound (nodal and numbers) and outbound coupled
Nodal monitor	from admission control	✓	✓	✓
	from outbound control	**X**	**X**	**X**
Number monitor	from admission control	**X**	✓	**X**
	from outbound control	**X**	**X**	✓

[*] Outbound may or may not be present.

Service logic may effectively split a traffic stream (for the same service and CTC) over multiple destinations. The manner in which this can occur should be taken into consideration when coupling outbound control to inbound. Load sharing over multiple destinations should ideally be done dynamically (e.g. by using static proportional distribution in combination with alternative routing on destination busy), so that a situation cannot arise where one branch is very busy while another is not. This avoids a service provider losing calls which might otherwise have been carried on less busy branches. If it is possible to construct service plans with static proportional distribution only, then it is possible that such losses could occur if a service provider does not monitor the effectiveness of their chosen traffic distribution proportions. Furthermore, if outbound is coupled to inbound control, additional losses may be incurred when inbound control becomes activated, because calls may be rejected by inbound control that could otherwise have been carried on the branches that are more lightly loaded.

The best way to avoid any such problem is to ensure that a good load balancing method is always incorporated into a service plan that uses proportional traffic distribution over multiple destinations. This is highly desirable anyway from a service provider's perspective for the reasons pointed out above.

8.5.3.3   Avoiding Bias for Televotes

Televoting types of service provide a list of telephone numbers for callers to dial, and the successfully terminating calls to each number are counted. Most often the list has a common leading digit substring, e.g. 0123456789 for the numbers {01234567890, 01234567891,..., 01234567899}, but this is not necessarily the case.

Sometimes a list of non-contiguous numbers, such as {01231111111,01232222222} is required.

Calls for distinct televote numbers must terminate on a common network destination resource (line group), because if separate resources were used bias would be introduced. Multiple random (Poisson) traffic streams to a line group each see the same probability of being blocked (for any holding time distribution) because a given call being blocked is independent of the stream to which it belongs. No bias is introduced for a common resource, because the televote counts are proportional to the probability of a call being accepted, which is the same for all streams. On the other hand, if separate terminating resources were used, the probability of blocking would depend upon the arrival rate for each voting stream and the number of lines. The effect of this could be to make the counts closer.

The restriction function used by the control will usually behave in a similar way to the terminating resource, and therefore, to avoid bias, it is necessary to control all televote streams together with a single instance of restriction. Automatic detection of called numbers and application of control cannot be used, because this would apply separate controls for each number and introduce bias. Therefore, it is necessary to be able to manually specify the numbers to be controlled for televoting. This does not change the restriction updating (including activation and termination) function, which must still be automatic.

The detection and monitoring function requires the following enhancements, in order to manually specify control on televote numbers. All televote call-failure indications need to be counted by a single monitor, in order that the televote numbers are controlled as a single group (with a single gap interval and timer at each SSP). The monitor is reserved, and not available for dynamic allocation. It is also necessary to prevent any monitors which match the televote and have a longer number length from being dynamically allocated. This is easily done by searching for manually allocated monitors first and, if a matching number is found, disallowing the searching of any monitors with longer string length.

The IN call gap capability allows a single gap control to apply to the leading digit substring of a number (see section 8.2.1), but it does not allow a single gap control to apply to a list of numbers (or substrings thereof). Therefore, it is not possible to properly avoid bias with non-contiguous lists of televote numbers, and providers of services that use IN infrastructure should be warned of the consequences of using them.

8.5.4   ACR Control

8.5.4.1   Adaptation

Control adaptation has two fundamentally opposing objectives:

- the control level should respond rapidly to changing offered calling rates;

- it should give stable admitted calling rates when the offered calling rate is constant.

Making a feedback control more responsive tends to make it less stable, and vice versa.

The ACR control has been applied with several different methods of control adaptation. In the IN context, coefficients that multiply the gap or leak interval at each change, are being used successfully. A distinct 'increase coefficient' (>1) and 'decrease coefficient' (<1) is used to increase or decrease the interval respectively. These coefficients themselves can be adapted to speed up convergence after a sudden change in load, while retaining stability under constant load.

The level of restriction to use when control is first activated can be made less arbitrary if the possible ways in which overload starts are considered (see section 8.6.2.2 for an example). Starting with the maximum possible required level of restriction under any condition is usually too severe. It is better to choose a lower (but still high) level that is related to the conditions under which overload onset is likely to occur suddenly. If a higher (or lower) level of restriction is required then the control can be relied upon to quickly adapt accordingly. For example, the call attempt rate will increase due to the effect of repeat attempts, which means that the level of restriction will have to increase from the initial value (see section 8.6.3).

Termination of control must be handled carefully, in order to prevent premature release of control. This can then lead to immediate re-activation, with a restriction level that is too high, because of the reasons above. The consequence of this is highly oscillatory control and reduced throughput.

Mechanisms for effective termination of control are discussed in section 7.6.3.5 of Chapter 7.

8.5.4.2   Avoidance of Synchronisation

Control updating, for restriction methods which use an interval timer such as gapping and leaky buckets (at high load), would, under certain conditions, suffer from traffic synchonisation, if the control were not designed to prevent the effect. It occurs when the interval timers at each point of restriction become synchronised. Similar effects have been demonstrated before in the context of switching systems [7, 8].

Synchronisation is more likely to arise under the following conditions:

- high calling rates;
- large numbers of sources (e.g. SSPs);
- sudden increase in offered rates.

The first two of these imply that a large interval will be used, and the last one can make them synchronised because, as soon as control is initiated, call gap messages are distributed rapidly to many of the SSPs. The result would be a very peaky (batched) arrival process.

For inbound control, synchronisation can be avoided by varying the gap interval sent to each SSP from where an IDP is received, after a given restriction level update. Modelling has shown that an effective method is to increase the gap interval over a range from below to above the value computed by the updating function, for each successive SSP. To do this, it is necessary to record when an SSP has been sent a message after each update.

This method is better than randomising the gap interval, which could destabilise control when there are only a small number of SSPs originating traffic.

8.6 Parameter Optimisation and Performance

8.6.1 Approximate Flow Model

Ignoring detailed stochastic characteristics of the load and developing an approximate model based upon mean calling rates provides useful insight into the behaviour of the controlled rates. Furthermore, under steady load, it gives good agreement with simulation.

8.6.1.1 Model for SCP Capacity Allocation (Controlled Admitted Rates)

In the absence of outbound control coupled to inbound control, it is possible to derive how the available SCP capacity, as determined by the SCP admission (internal) control function, will be partitioned among the streams of calls to distinct numbers. This depends only upon the controlled target reject rates and admission control function, and not on the particular method of call restriction.

The notation used will be as follows.

- At the gapping function (SSPs):

  Λ IDP (offered) calling rate, from all SSPs;

  λ rate admitted by gapping (offered to SCP admission control);

  B probability of being rejected downstream, including gapping or SCP admission control.

- At SCP admission control:

  γ admitted rate;

  γ_R admitted rate when the rate rejected by admission control is R;

  b probability of a call being rejected (blocked) — it is assumed that the probability of a call being rejected is the same for all streams, which is reasonable since most admission control schemes do not differentiate between different call priorities (apart from emergency calls).

Indexes applied to each rate indicate either stream identity (by dialled number or address) or SSP identity, depending on the context.

An SCP nodal monitor bucket will be assumed to have a leak rate of R, and monitors that can be assigned to specific number stream $i(1 \leq i)$ will have a leak rate of r_i, where:

$$R \geq \sum_{i \geq 1} r_i$$

Nodal (Inbound) Control Only

The simplest case is when nodal control only is present (as defined by a single SK), i.e. there is a single monitor for all rejected calls, with a leak rate R say.

When the ICF gapping control is active and steady it will maintain the rate at which calls are rejected by SCP admission control A to be the leak rate R. With the assumption that the reject rate of A increases as the rate it is offered increases (see section 8A.1.1 of the Appendix to this chapter), there will be a unique admitted rate γ_R (which can vary if the admission control depends upon the call mix) when the reject rate is R. Therefore, for all streams, the probability that a call is admitted, by gapping and A, is:

$$\frac{\gamma_R}{\Lambda}$$

and the rate admitted by each stream i is:

$$\Lambda_i\left(\frac{\gamma_R}{\Lambda}\right) = \left(\frac{\Lambda_i}{\Lambda}\right)\gamma_R$$

That is, each stream gets an admitted rate proportional to its offered rate.

Nodal and Specific Number (Inbound) Control (No Outbound Control)

Now suppose that there is a set of monitors for number-specific control as well as the nodal monitor, and rejects are registered by both (assuming sufficient are available). Below we will present a solution to the partitioning of capacity, but an outline of how to prove that the solution is unique and how many streams are under control is given in this chapter's Appendix (section 8A.1.2). To show that convergence to the solution is stable requires modelling the system with ordinary differential equations. For this, the reader is referred to section 7.6.4 in Chapter 7, where the mean calling rates are dependent upon time. Adaptation of the admitted rate towards the target reject rate is expressed in terms of the rate of change of the admitted rate.

The Appendix (8A.1.2) shows how to determine the number of streams under control. Let the set of leak rates (smaller than R) for each stream under control be $\{r_i\}$. The Appendix (8A.1.3) also shows that the total admitted rate is partitioned into admitted rates for streams of specific called numbers $\{\gamma_i\}_{i \geq 1}$ where:

$$\gamma_i = \left(\frac{r_i}{R}\right)\gamma_R, \quad i \geq 1$$

The remaining background admitted rate γ_0 under nodal control is made up of calls to a range of non-specific numbers:

$$\gamma_0 = \left(1 - \sum_{i \geq 1}\frac{r_i}{R}\right)\gamma_R$$

From this we see that the relative capacity for sufficiently large number-specific streams are in proportion to their monitor leak rates. The background traffic, made up of streams that are not large enough to be controlled individually, gets the remaining capacity. This is useful, because it means that the SCP capacity can be divided among overload streams by setting monitor leak rates. Furthermore, setting:

$$R > \sum_{i \geq 1} r_i$$

would ensure that low rate 'background' streams do not get pushed out altogether when the maximum number of streams are under control. In particular, if there are a maximum of n complete number monitors each of equal leak rate r (a likely configuration), then they each get r/R of the available capacity, while the remaining traffic gets the proportion $1 - nr/R$. Note that this means that no particular stream can dominate the available capacity, which can occur when there is a single nodal monitor.

However, given that the SCP has no knowledge of the rate at which connections to the network are ineffective (do not terminate), they can both be regarded as fair. A single nodal monitor is fair in the sense that all streams see the same probability of being rejected by the SCP (but not by the subsequent network connection attempt). The addition of a set of monitors for number-specific control is fair in the sense that the SCP capacity is divided up equally among each of the streams. What is a correct interpretation of fairness is debatable (see sections 7.5.6 and 7.6.5 in Chapter 7 for further discussion).

More critical than the issue of fairness is that, without outbound control as well, coupled to inbound control (see section 8.5.3.2), it can be shown that two problems can occur:

- over-restriction of some streams, i.e. network destinations may be starved of enough traffic to ensure that they are highly utilised;

- under-restriction of some streams, which implies higher levels of ineffective traffic in the SCP, and hence other streams experience a higher loss than is necessary (when nodal control is active).

Nodal and Specific Number Inbound and Outbound Control (Coupled)

If outbound control is present and coupled to inbound, as in the last column of Table 1, then it can be shown to have the following effect. All traffic streams will see the same probability of blocking due to SCP admission control. After this thinning, any streams to specific called numbers that are sufficiently large are limited by outbound control to the network terminating capacity (with reject rate the target monitor reject rate). This chapter's Appendix (8A.2.1) demonstrates how to obtain a solution for the admitted rate, and which streams are under control. If the controlled rates for these are $\{c_i\}$, then the remaining streams get the following admitted rate (if under inbound control) (see section 8A.2.2):

$$\gamma_0 = \gamma_R - \sum_1^n c_i$$

As long as the factors noted in section 8.5.3.2 are taken into account when coupling inbound to outbound control, then this represents ideal behaviour. It can be summarised as follows: each stream is equally likely to be rejected due to limited SCP nodal capacity, but cannot grow bigger than the network terminating capacity will allow.

8.6.1.2 *Call Gapping (Inbound Control) Behaviour*

The analysis of capacity allocation in section 8.6.1.1 showed that, for a given stream that is under control, there is a precise controlled mean rate at which calls are offered to SCP admission control (i.e. the rate after gapping). Adaptation of the gap interval drives convergence to this rate.

For a particular stream of calls under control, let there be random call (IDP) arrivals at SSP i with rate Λ_i, and a gap interval of τ. By considering the mean time between admitted calls, it can be seen that the total mean admitted rate from all N SSPs is given by:

$$\lambda = \sum_{i=1}^{N} \lambda_i \quad \text{where} \quad \lambda_i = \frac{1}{\tau + 1/\Lambda_i} = \frac{\Lambda_i}{\tau \Lambda_i + 1}$$

We can deduce some useful properties of this admittance function.

- For a given total fixed offered rate from all N SSPs and given gap interval, the maximum admitted rate after gapping occurs when the offered rates from each SSP are the same. This can be shown by solving:

$$\text{maximise } \lambda = \sum_{1}^{N} \frac{1}{\tau + 1/\Lambda_i}$$

$$\text{subject to } \sum_{1}^{N} \Lambda_i = \Lambda ,$$

e.g. using Lagrange multipliers. Conversely, the minimum admitted rate occurs when all the traffic Λ is generated at the minimum number of SSPs (and 0 elsewhere). These results are used to derive the minimum, maximum, and initial gap interval.

- The rate of change of the total rate admitted by gapping with respect to the gap interval is:

$$\frac{\partial \lambda}{\partial \tau} = -\sum \lambda_i^2 .$$

For a fixed admitted rate $1 = \gamma_R + R$, it can also be shown that this rate of change is minimum when traffic is evenly distributed over the SSPs, and the maximum occurs when the traffic originates from just one SSP. In fact, the bounds are:

$$\frac{\lambda^2}{N} \leq \left| \frac{\partial \lambda}{\partial \tau} \right| \leq \lambda^2 .$$

This is important for designing the control updating strategy because it shows that changes to the gap interval produce a greater change to the admitted rate when the traffic is unevenly distributed over SSPs.

8.6.2   Optimisation of Parameter Values

Whatever the precise details of ACR implementation, it will have a set of parameters that must be assigned values that make the control work effectively. A list of the type of parameters is presented in Table 8.2.

Table 8.2   Control parameter types.

Detection and Monitoring	Restriction updating
Monitor fill thresholds and leak rate	Update timer length parameters
Number of monitors for each number string length	Initial, maximum, minimum levels of restriction
	Increase and decrease parameters

These are required for both the ICF and OCF, which may use different restriction control algorithms. It would not be appropriate to give specific examples of every parameter and show how to derive optimal values for each in this chapter. Instead, a flavour for what is involved is given by outlining below methods for deriving the update timer length and the initial gap interval for the ICF.

8.6.2.1   *Update Timer Length (for the ICF)*

The update timer is necessary whenever the monitor remains in the same state, i.e. a high or low state, to increase or decrease the level of restriction respectively. It is reset after each expiry. There are two opposing constraints to choosing the value of the update time. It must be:

- short enough to enable fast response to changing traffic rates (responsiveness);
- long enough so that sufficient control (gap) messages have been distributed and that the mean arrival rate at the SCP has converged close enough to the limiting value implied by the most recently distributed gap interval (stability).

There are two aspects to deriving a sufficiently long timer length. One is to analyse how long it takes to distribute call gap messages. This alone is not entirely satisfactory, because when the calling rate is low it takes a longer time to distribute them, although the effect on the admitted call rate may be quite small. Therefore, it is better to also estimate this effect as well. The approach is rather involved, and so will only be outlined.

To understand the distribution of call gap messages, an event-timing diagram is helpful. Figure 8.6 shows how gap messages are distributed after an update to the gap interval is made at time 0, when the SSP-SCP delay is d_1 and that in the opposite direction is d_2, with round-trip time $d_1 + d_2 = D$. D is assumed constant since we are concerned with the upper limits of the round-trip time, rather than trying to take account of its variability. Suppose an IDP arrives at the SCP at time t (say) after the update was made, which was therefore sent from an SSP at time $t-d_1$. The call gap message is returned, arriving back at the SSP time d_2 later. The new gap interval value τ_1 can only be applied after this call gap message is received and any already running gap timer of length τ_0 has expired.

With reference to Fig 8.6, it can be seen that the earliest time that the new interval can take effect is $t + \tau_0$ at the SCP when $\tau_0 > D$, and $t + D$ otherwise. The latter occurs when τ_0 is just less than D and an interval timer is started just before the call gap message arrives.

Now t is a random variable. On the simplifying assumption that the system of gap interval timers running over all SSPs is in the steady state, i.e. there is independence between the phasing of gap timers between SSPs, it is possible to derive its distribution (see section 8A.3 in the Appendix to this chapter). This gives an expression for:

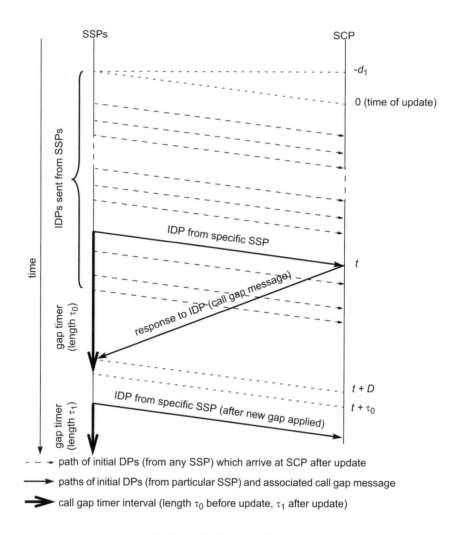

Fig 8.6   Call gap distribution.

$p(t) = \Pr\{$an IDP from any given SSP arrives at the SSP in the interval $[0,t]\}$

Let λ_0 and λ_1 be the total mean (long-term) admitted rates from all SSPs after gapping with intervals τ_0 and τ_1 respectively (calculated according to section 8.6.1.2). (Note that in this context the indices no longer refer to SSPs or streams.) If there are N SSPs with equal offered rates, then it can be shown (see section 8A.4) that the mean rate $\lambda(.)$ at the SCP as a function of $t+\tau_0$ is given by:

$$\lambda(t + \tau_0) = N \cdot (\lambda_0 + (\lambda_1 - \lambda_0)p(t))$$

To find the minimum length of time it is necessary to wait in order for the mean admitted rate to be within Δ of the limiting rate $N\lambda_1$, it is therefore necessary to solve (for t):

$$N \cdot (1 - p(t)) |\lambda_0 - \lambda_1| = \Delta$$

This enables an estimate to be obtained of the minimum value of the update time that should be applied, in terms of the current value of the gap interval (when it is greater than the round-trip time D). When the gap interval is smaller than D, D gives the lower bound.

8.6.2.2   *Initial Nodal Gap Interval (for the ICF)*

Nodal control could be activated as a result of a gradual rise in the level of offered traffic above the SCP capacity, but this is unlikely to be a common occurrence. If it does occur, then it is not difficult to control because of the slowly changing traffic level.

Much more likely is the need to control background traffic as a result of the sudden increase in the calling rate, due either to re-routed traffic following an SCP node failure, or to a sudden rise in the traffic to a specific called number. It will be assumed that the node failure case is worse since it can happen virtually instantaneously. This scenario is therefore used to determine the initial gap interval for nodal control.

For number-specific control, an alternative scenario based upon a media-stimulated event would be used to determine the initial gap interval (not presented here).

For the purposes of studying the effects of SCP failure, it will be assumed that traffic originates uniformly from all SSPs. This is a reasonable assumption given that the overloading traffic is not due to a specific number, and that the network is designed to evenly load SCPs under such circumstances. Furthermore, it leads to the highest gap interval (see section 8.6.1.2).

We will assume that there are three SCPs and traffic is alternatively routed under failure as in section 8.1.2. The effects of SCP failure are worst when, just before failure, the SCPs are loaded to the maximum possible before control is activated. Although this apparent worst-case is used to determine the initial gap interval for nodal control, nonetheless, if in reality this proves too great, control would rapidly decrease the interval to the appropriate value. It is deemed more important to quickly reduce the traffic arrival rate for sudden overload onset, than to avoid losing a few additional calls while the control relaxes due to initial gapping being too excessive.

In fact, the gap interval derived here is not the maximum possible, which occurs with an infinite arrival rate originating from the greatest number of SSPs. This turns out to be 3 times the initial interval (for 1 SCP failure). The value derived here ignores the effect of repeat attempts, which we will see in section 8.6.3 means that the gap interval must adapt upwards substantially. But it is reasonable to ignore repeats for the initial interval, since they build up progressively after the initial onset of overload, dependent upon the inter-repeat attempt time and probability. Let there be N SSPs per SCP. After only one SCP failure the number of source SSPs per site is $N + N/2 = (3/2)N$, and the total traffic offered just before failure is $3(\gamma_R + R)$. The total traffic offered per SCP just after failure is $3(\gamma_R + R)/2$, and therefore the required gap interval is:

$$\tau_0 = \frac{3}{2}N\left(\frac{1}{\gamma_R + R} - \frac{1}{\frac{3}{2}(\gamma_R + R)}\right) = \frac{1}{2}\frac{N}{(\gamma_R + R)}$$

Compare this with the maximum possible gap interval required when one SCP has failed, which is given by assuming an infinite arrival rate uniformly distributed over all SSPs:

$$\tau_{max} = \frac{3}{2}N\left(\frac{1}{\gamma_R + R}\right) = 3\tau_0$$

8.6.3   Simulation Results for the Case of SCP Node Failure

A discrete event simulation has been developed to enable studies of control behaviour under different overload conditions. The example of SCP node failure presented above (section 8.1.2) is a good example to demonstrate how a control, based upon converging to a limiting call reject rate, behaves with an effective updating algorithm and good choices of parameter values. Outbound control is not relevant in this scenario and this considerably simplifies the complexity of control and its analysis. In particular, it is only necessary to consider a single stream (of all calls), not separate streams to different numbers, and enables us to focus on the dynamics of control.

It is important to include the effects of calling customer behaviour in the model, as explained in more detail in Chapter 7. Although a customer has an intent to make a (successful) call, a particular attempt may fail, in which case they may retry. The failure may occur because the attempt has been explicitly rejected or because the caller has abandoned the call due to a slow response from the network. The latter should only occur when no overload control is present or the control is not working properly. In any case, the result is that the call attempt rate is usually greater than the intent rate, which can be estimated when the persistence probability is known.

In this illustrative example, for each of two SCP nodes, call intents are generated randomly at the multitude of its associated SSPs. The total rate from all SSPs is just below the SCP's capacity (see Fig 8.7), and therefore the reject rate is insignificant and thus call gapping is not active. At time 50 sec, one of the SCPs fails and re-

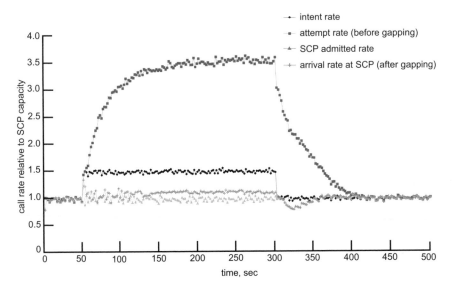

Fig 8.7   Simulation results for the case of SCP node failure:calling rates.

routes half of its traffic on to the remaining working SCP, which activates nodal control (gapping) as a result of the increased reject rate. Nodal control quickly adapts the gap interval so that the mean reject rate has the value used for the nodal monitor leak rate (Fig 8.8).

The repeat attempts increase so that the call attempt rate increases to over 3 times the intent rate, due to a persistence probability of 0.8. The gap interval has to increase from its initial value of 0.113 sec to account for this, eventually settling down at about 0.195 sec.

Finally, the failed SCP starts working again at time 300 sec, and the control gradually turns off. At time 399 sec, the gap interval reaches the minimum (12 ms) and the SCP starts to rapidly turn off gapping at the SSPs by distributing messages with a gap duration set to 0. It is not the state of the SSP gapping shown in Fig 8.8, but that of the SCP control. This remains at the minimum gap interval for more than 100 sec, in case the offered rate should increase again, before terminating control at the SCP.

The benefit of this behaviour is that the control quickly activates to maintain a maximum rate admitted by the SCP. Although it has not been shown here, this will bound the SCP response time (for any load originating at the SSPs). It also maximises the throughput of the SCP by limiting the ineffective calls due to SCP rejection, subject to this response time requirement. These requirements were stipulated in section 8.4.1.

We will now show that these simulation results actually show very good agreement with theory along the lines of that presented in section 8.6.1.

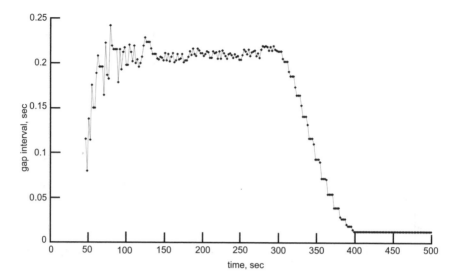

Fig 8.8   Gap interval adaptation for the case of traffic re-routing under SCP node failure.

If the total intent rate is $\hat{\Lambda}$, then a simple model (see section 7.3.1 of Chapter 7) implies that the attempt rate, which includes repeats with probability p, is the intent rate inflated:

$$\Lambda = \frac{\hat{\Lambda}}{1 - Bp}$$

where B is the probability that a call is rejected by the network. For even traffic origination from N SSPs, the gap interval required is given by:

$$\tau = N \left(\frac{1}{\lambda_R} - \frac{1}{\Lambda} \right)$$

where λ is the rate admitted by gapping and offered to the SCP. If γ_R is the maximum rate admitted by SCP admission control, then the following equation relates this to the call attempt rate and the probability of being rejected (by gapping or the SCP):

$$\gamma_R = \Lambda (1 - B)$$

and the following expresses the condition that nodal control is active:

$$\lambda_R = R + \gamma_R$$

These together imply that the required gap interval is given by:

$$\tau = \left(\frac{\gamma_R}{N}\right)^{-1} \left(\frac{1}{R/\gamma_R + 1} - \frac{(1-p)}{\hat{\Lambda}/\gamma_R - p}\right)$$

In this last expression, the calling rates are given relative to the SCP capacity γ_R. For the case here the following values were used:

$$\hat{\Lambda}/\gamma_R = 1.512 \quad R/\gamma_R = 0.1233 \quad p = 0.8 \quad \gamma_R/N = 2.918$$

The required gap interval τ is 0.209 sec, giving very good agreement with the value to which the control converges in Fig 8.8. The limiting ratio of attempt rate to SCP capacity rate Λ/γ_R is 3.57, again showing good agreement with Fig 8.7.

8.7 Summary

The nature of typical IN traffic and the centralised (physical) IN architecture which is commonly deployed imply that effective IN overload controls are essential. IN standards are limited to specifying an information flow and call restriction feature, the call gap capability. There is therefore a need for significant development in order to obtain complete and effective IN overload controls.

To work well under a wide range of conditions, such as varying distribution of traffic over SSPs, or different relative sizes of traffic streams identified by different called (or calling) numbers, such overload controls must be adaptive. Here we have presented an adaptive control design that is based upon convergence to specified call failure rates. This has been applied to control incoming traffic to an SCP, where it is assumed that the SCP already has a call admission control scheme, and to traffic outgoing to the network. Approximate steady-state analysis, which agrees well with simulation, shows how the SCP capacity is partitioned with respect to traffic streams. This reveals that number-specific control of inbound streams ensures that no traffic stream can dominate and possibly cause high levels of ineffective traffic. However, the rate at which calls are successfully terminated in the network can be maximised by coupling outbound to inbound controls. Thus, the control cannot only ensure that response time requirements are met, but it can also maximise throughput by limiting ineffective calls.

There are many control parameters in such a scheme. For good control performance, optimal values of these parameters should be used. It has been shown here how values should be assigned to a subset of the parameters — the control update time, and the initial gap interval for control inbound to the SCP.

Many of the control features described in this chapter have been applied to BT's IN platforms. But the type of method applied here is not specific to INs. Networks based upon more recent and future technologies will undoubtedly be susceptible to similar problems once deployed with sufficiently large capacity and connectivity. It is anticipated that similar techniques to those proposed here will be applied to such technologies, e.g. the session initiation protocol (SIP) for multimedia call control, which is derived from the Internet Engineering Task Force (IETF).

Appendix

8A   Capacity Allocation

In this Appendix we will demonstrate how the SCP capacity (admitted rate) is partitioned among traffic streams in the steady state, and its dependence upon control monitor leak rates. This is done for the case of number-specific inbound control, and outbound control coupled to inbound control. It shows that the latter approach maximises effective traffic.

8A.1   Capacity Allocation for Inbound Control Uncoupled with Outbound Controls

8A.1.1   Characteristics of an Admission Control Function

We first need some preliminary results concerning the behaviour of the SCP admission control function (independent of any adaptive inbound call gapping controls) expressed in terms of its probability of blocking (rejecting) a call, b.

Recall (from section 8.6.2) that the rate offered to and admitted by A is denoted by λ and γ respectively. In addition, let the rejected (overflow) rate be ω. By definition:

$$b = \frac{\omega}{\lambda} \text{ and } \omega = \lambda - \gamma.$$

An admission control is assumed to be characterised by an admitted rate function (implicitly assuming that any hysteresis effects, due to different behaviour when the offered rate is rising from falling, are small) and usually has the following characteristics:

$$0 \leq \gamma \leq \lambda$$
$$\gamma'' < 0 \quad \text{for all } \lambda > 0$$

where $'$ denotes differentiation with respect to λ. The latter condition follows from the need to reject more calls as the offered rate increases, which takes some effort and hence reduces the available capacity.

Although γ and ω are functions of λ they can also be written as functions of b because $b' > 0$ when $\lambda > 0$. This follows because:

$$b' = \frac{1}{\lambda^2}(\gamma - \gamma'\lambda), \text{ and } \gamma - \gamma'\lambda,$$

which takes the value 0 at $\lambda = 0$, has derivative $-\gamma''\lambda > 0$ in view of the admission control characteristics.

$\dfrac{d\omega}{db}$ and $\dfrac{d^2\omega}{db^2}$ are therefore well-defined.

$\dfrac{d\omega}{db} = \lambda\left(1 + \left(\dfrac{\lambda - \gamma}{\gamma - \gamma'\lambda}\right)\right)$ is also always positive.

It can also be shown that, for admission control functions of interest, $\dfrac{d^2\gamma}{db^2}$ is also positive.

8A.1.2 Existence and Uniqueness of the Solution

Let $\{\Lambda_i\}_{i=1...n}$ be the set of all originating calling rates (at SSPs) of streams to distinct numbers which satisfy $\Lambda_i \geq r_i$, where r_i is the monitor leak rate for stream i. These traffic streams may be subject to number specific control. Let Λ_0 be the rates of all remaining streams, which cannot be subject to number-specific control because their individual rates are too small. The rate at which stream i calls are rejected by SCP admission control is $\omega_i(b)$. In the steady-state the gapping control at the SSPs limits this to the monitor leak rate:

$$\omega_i(b) \leq r_i$$

In view of this, define the following overflow rate functions of the admission control blocking probability b:

$$\phi_0(b) = \Lambda_0 b$$

$$\phi_i(b) = \begin{cases} \Lambda_i b & b < b_i \\ r_i & b \geq b_i \end{cases} \quad \text{for } 1 \leq i \leq n$$

$$\text{where } b_i = r_i/\Lambda_i$$

At this point, we assume that nodal control is not active, so that the originating call attempts rates $\{\Lambda_i\}$ can be used, rather than those after gapping, $\{\lambda_i\}$. Note that each $\phi_i(b)$ is bounded, because at the SCP reject probability of b_i the SSP gapping on this stream starts to take effect in order to limit the SCP reject rate. Without loss of generality assume that the indexing of the streams is such that $b_n \le b_{n-1} \le \dots \le b_1$. Now consider the total reject rate at the SCP, given by:

$$\phi(b) \equiv \sum_{i=1}^{n} \phi_i(b) = \begin{cases} \displaystyle\sum_{0}^{n} \Lambda_i b & 0 \le b < b_n \\[2em] \displaystyle\sum_{0}^{k} \Lambda_i b + \sum_{k+1}^{n} r_i & b_{k+1} \le b < b_k \quad (1 \le k < n-1) \\[2em] \Lambda_0 b + \displaystyle\sum_{1}^{n} r_i & b_1 \le 1 \end{cases}$$

This function is piecewise linear, continuous, and increasing, with decreasing change of gradient at the 'corners', and is illustrated in Fig 8A.1.

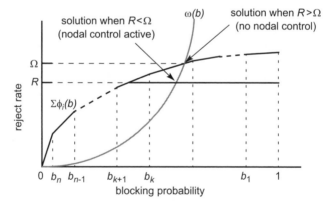

Fig 8A.1   Reject rates for inbound control only when several controls may be active simultaneously.

The nodal monitor leak rate R places an upper limit on the total nodal reject rate:

$$\sum_{0}^{n} \omega_i(b) \le R$$

Since $\phi(b)$ is still valid for the region satisfying this inequality, the admissible form of the total reject rate is given by the joining of the above piecewise linear curve with the line due to this constant upper limit. This is also illustrated in Fig 8A.1.

The reject rate function $\omega(b)$ due to SCP admission control is also shown. For this curve, the total arrival rate (after gapping) increases with increasing b, whereas for the piecewise linear curve the offered arrival rate is constant. Because we are assuming that:

$$\frac{d^2\omega}{db^2} > 0,$$

it follows that there is a unique solution for the blocking b at the intersection of these curves with reject rate Ω, say. If $R < \Omega$, then nodal control is active, otherwise it is not. If the intersection of the line of equal rate R or Ω (whichever is less) is between b_{k+1} and b_k, then there are $n-k$ specific numbers under control in addition to nodal control.

8A.1.3   Admitted Rate Partitioning when the Number of Streams Under Control is Known

Section 8A.1.2 showed the existence and uniqueness of the solution in the steady state, in terms of the SCP admission control blocking probability, the specific called number streams under control, and whether nodal control is active. We now show how the SCP capacity (i.e. the admitted rate) partitions over each stream, assuming that the number of streams under control is known, and nodal control is active.

For simplicity we now change the interpretation of the notation slightly, and index 0 is applied to all streams that are not under number-specific control ('background' streams). The total rate λ offered to A is partitioned into:

$$\lambda = \sum_{i \geq 0} \lambda_i$$

Given the function A, there is a unique offered rate λ_R and admitted rate γ_R at which the reject rate is R (see section 8A.1.2). The rejection probability at A can be expressed as:

$$b = \frac{R}{\gamma_R + R} = \left(1 + \frac{\gamma_R}{R}\right)^{-1}$$

The reject overflow (reject) rate of each (controlled) stream i is:

$$\omega_i = \lambda_i b \qquad i \geq 1$$

and, under control, this is equal to the full number length monitor leak rate, that is:

$$\omega_i = r_i$$

Therefore the offered rate for stream i is:

$$\lambda_i = r_i/b$$

It follows that the admitted rate for each stream i is (as presented in section 8.6.1.1):

$$\gamma_i = \lambda_i(1-b) = r_i\left(\frac{1-b}{b}\right) = \gamma_R\left(\frac{r_i}{R}\right), \qquad i \geq 1$$

The global monitor leak rate is the sum of the leak rates for all number-specific streams under control plus the 'background' rate:

$$R = \omega_0 + \sum_{i \geq 1} \omega_i = \omega_0 + \sum_{i \geq 1} r_i$$

Since $R - \sum_{i \geq 1} r_i = \omega_0 = b\lambda_0$, it follows that the admitted rate of background calls is

(as presented in section 8.6.1.1): $\gamma_0 = \lambda_0(1-b) = \left(R - \sum_{i \geq 1} r_i\right)\left(\frac{1-b}{b}\right) = \left(1 - \sum_{i \geq 1} \frac{r_i}{R}\right)\gamma_R$.

8A.2   Capacity Allocation for Coupled Controls

8A.2.1   Existence of Solution

Now let $\{\Lambda_i\}_{i=1...n}$ be the set of all originating calling rates (at SSPs) of streams to distinct numbers which satisfy $\Lambda_i \geq c_i$ where each outbound control for stream i has (maximum) controlled rate c_i (including outbound rejects). These may be subject to number-specific control. c_i is the sum of the rate at which calls are successfully terminated by the network destination and the associated ICF monitor leak rate. Note that inbound gapping on such streams will only be activated once these rates have been attained.

As before, let Λ_0 be the rates of all remaining streams, which cannot be subject to number-specific control because their individual rates are too small.

The behaviour of number-specific control applies the following constraints:

$$\gamma_i \leq c_i \qquad 1 \leq i \leq n$$

where γ_i is the rate admitted by inbound admission control for stream i, (for which control is active only when this is an equality). Define the following admitted rate functions of the SCP node admission control blocking probability b:

$$\mu_0(b) = \Lambda_0(1-b)$$

$$\mu_i(b) = \begin{cases} \Lambda_i(1-b) & b > b_i \\ c_i & b \le b_i \end{cases} \quad \text{for } 1 \le i \le n$$

where $b_i = 1 - c_i / \Lambda_i$

Without loss of generality, assume that the indexing of the streams is such that $b_1 \le b_2 \le \ldots \le b_n$ (the reverse order of section 8A.1.2). The total admitted rate is therefore given by:

$$\mu(b) \equiv \sum_{i \ge 0} \mu_i(b) = \begin{cases} \Lambda_0(1-b) + \sum_1^n c_i & 0 \le b < b_1 \\[2ex] \sum_0^k \Lambda_i(1-b) + \sum_{k+1}^n c_i & b_k \le b < b_{k+1} \quad (1 \le k \le n-1) \\[2ex] \sum_0^n \Lambda_i(1-b) & b_n \le b < 1 \end{cases}$$

It follows that the function is piecewise linear, continuous, and decreasing, with increasing change of gradient at the 'corners', as illustrated in Fig 8A.2.

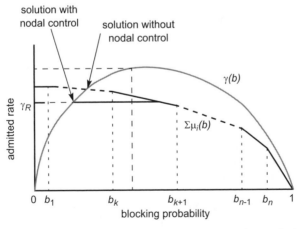

Fig 8A.2   Admitted rates for coupled inbound and outbound control when multiple controls may be active simultaneously.

Inbound admission control implies a maximum admitted rate γ_R given the nodal monitor leak rate R:

$$\gamma = \sum_{0}^{n} \gamma_i \leq \gamma_R$$

The section of the piecewise linear curve that is below the line $\gamma = \gamma_R$ is combined with this line, to obtain another piecewise linear curve. A solution is given by the intersection of this with the admitted rate function $\gamma(b)$. Figure 8A.2 illustrates this in the case when nodal control and $n - k$ outbound controls are active.

One can show that: $\dfrac{d\gamma}{db} = \dfrac{\lambda^2 \gamma'}{(\gamma - \gamma'\lambda)}$, which has the same sign as γ' because we showed in section 8A.1.1 that the denominator is positive. γ' may always be positive, or may become negative so that the admitted rate decreases to 0 as illustrated. The latter will occur in practice if the admission control function reflects the fact that rejection of calls always takes some positive processing effort, and hence the system may use up all of its effort rejecting demand. In that case, let the maximum possible rate offered to admission control be λ_{max}, which occurs when $b = 1$ (and $\gamma = 0$). The above expression for the gradient of γ at this point evaluates to give:

$$\lambda_{max} = -\frac{d\gamma}{db}\bigg|_{b=1}$$

The piecewise linear curve has gradient $\sum_i \Lambda_i$ at $b = 1$, so that it can then be seen that it will be below $\gamma(b)$ if $\sum_i \Lambda_i < \lambda_{max}$, and above otherwise.

More generally, it can be seen that lower offered rates lower this piecewise continuous curve and decrease the $\{b_i\}$ values, and raising offered rates produces the converse effect. High offered rates could lead to nodal control only, where γ_R is attained before any outbound control is activated. There could be more than one intersection of these two curves, but this cannot happen if the blocking probability, at which the reject rate is R, is at the maximum of $\gamma(b)$ or less than this (see Fig 8A.2).

8A.2.2   Admitted Rate Partitioning when the Number of Streams Under Control is Known

The analysis of capacity partitioning when the streams under control (including nodal) are known is simple. It is convenient as before to change the indexing so that 0 refers to the amalgamation of all streams under nodal control, and 1 to n refers to the streams under outbound control. If the controlled rates for the latter are $\{c_i\}$, then clearly the remaining streams get the following admitted rate:

$$\gamma_0 = \gamma_R - \sum_1^n c_i$$

8A.3 Distribution of Time to Send an IDP after a Gap Interval Update

Let the update time at the SCP be 0, corresponding to $-d_1$ at SSPs. For any particular SSP let T be the random variable that is the time after time $-d_1$ to sending the next IDP and τ be the length of the gap interval. Let S be the random variable that represents the length of time for which the gap timer has been running, and let the rate at which IDPs are offered to the SSP be Λ. Assuming that gapping is in the steady state, then S is uniformly distributed, and so:

$$\text{Pr}\{\text{interval timer is now running}\} = \frac{1}{1 + \Lambda\tau}$$

and

$$F(s) = \text{Pr}\{S < s\} = \frac{s}{\tau}\left(\frac{\Lambda\tau}{1+\Lambda\tau}\right)$$

Then:

$$\text{Pr}\{T < t\} = \text{Pr}\{T < t \,|\, \text{interval timer is not running}\} \cdot \text{Pr}\{\text{interval timer is not running}\}$$

$$+ \int_{s = \max\{0,\, \tau - t\}}^{s = \tau} \text{Pr}\{T < t \,|\, \text{interval timer has run for } s\} \cdot \frac{dF}{ds}\, ds$$

Therefore:

$$p(t) = \text{Pr}\{T, t\} = \frac{1}{1 + \Lambda\tau}(1 - e^{-\Lambda t}) + \left(\frac{\Lambda}{1+\Lambda\tau}\right) \int_{s = \max\{0,\, \tau - t\}}^{s = \tau} [1 - e^{-\Lambda(t - (\tau - s))}]\, ds$$

$$= \begin{cases} \dfrac{t\Lambda}{1 + \Lambda\tau} & \tau \geq t \\[4mm] 1 - \dfrac{e^{-\Lambda(t - \tau)}}{1 + \Lambda\tau} & \tau \leq t \end{cases}$$

8A.4   Mean Total Admitted Rate after Gapping

Let $p(t) = \text{Pr}$ {a call is sent from any given SSP in the time interval $[-d_1, t-d_1]$} as obtained in section 8A.3. Assume that the offered calling rate at each SSP is the same, and that call gaps have been fully distributed to each SSP since the last update so that the mean admitted rate from each SSP is the same (λ_0). Because the traffic streams from each source are independent, then:

$$p_k(t) = \text{Pr} \{k \text{ source SSPs have sent an IDP in time interval } [-d_1, t-d_1]\}$$

has a binomial distribution.

Let the (long-term) admitted rate from each SSP after reception of the new gap interval be λ_1. The total expected calling rate from all SSPs arriving at the SCP at time $t+\tau_0$ is therefore:

$$\lambda(t+\tau_0) = \sum_{k=1}^{k=N} E[\text{calling rate}|k \text{ sources have sent an IDP at time } t-d_1] \cdot p_k(t)$$

$$= \sum_{k=1}^{k=N} ((N-k)\lambda_0 + k\lambda_1) \cdot p_k(t)$$

$$= N(\lambda_1 + (\lambda_1 - \lambda_0)p(k)) \qquad \text{(using binomial property)}$$

References

1   ITU-T, Q.1214: Distributed Functional Plane for Intelligent Network CS-1, section 5.4.2 Call Gap Capability, section 6.4.2.9 Call Gap (1995); Q.1224: Distributed Functional Plane for Intelligent Network Capability Set 2, section 11.4.2 Call gap capability, section 12.4.3.12 Call Gap (1997).

2   Abernethy, T. W. and Munday, A. C.: '*Intelligent networks, standards, and services*', BT Technol J, **13**(2), pp 9-20 (April 1995).

3   '*A method of controlling overload in a telecommunications network*', Priority application country and number: EP 93309185.2; Priority application date (18 Nov 1993).

4   ITU-T, E.412: Network Management Controls (1998).

5   Langlois, F. and Régnier, J.: '*Dynamic congestion control in circuit-switched telecommunications networks*', Teletraffic and datatraffic in a period of change, ITC-13, Elsevier Science Publishers B V (North Holland) (1991).

6   Williams, P.: '*A novel automatic call restriction scheme for control of focused overloads*', 11th UK IEE Teletraffic Symposium, Cambridge (1994).

7   Erramilli, A. and Forys, L.: '*Oscillations and chaos in a flow model of a switching system*', IEEE Journal on Selected Areas in Communications, **9**(2) (February 1991).

8   Nyberg, C., Wallstrom, B. and Korner, U.: '*Dynamical effects in control systems*', IEEE Telecommunication Traffic, **2**, pp 778-781 (December 1992).

9

CAPACITY PLANNING FOR CARRIER-SCALE IP NETWORKS

C A van Eijl

9.1 Introduction

Operators of carrier-scale IP networks are faced with the constant challenge of providing a consistent service in the most cost-effective way while dealing with rapid network expansion and traffic growth, the introduction of new services, and the deployment of new technologies.

This is particularly challenging for networks that offer quality of service (QoS). Such networks support a wide range of application with very different requirements on performance. They enable customers to prioritise their applications according to performance and business need by offering multiple classes of service with different targets on delay and throughput, backed by service level agreements (SLAs) or service level guarantees (SLGs). For example, voice applications should be assigned to a high priority class to ensure the low delay and loss that is necessary for a satisfactory call quality. On the other hand, Web browsing and e-mail are usually considered to be lower priority and tolerate a 'best-effort' service.

Since businesses are increasingly moving towards one IP-based platform for all their applications, there is a growing demand for IP quality of service, particularly in enterprise networks and IP virtual private networks (VPNs). The latter provide the customer with a highly flexible and secure corporate IP network between geographically dispersed sites via a shared and managed infrastructure.

The underlying mechanisms that enable service differentiation in an IP network gave already been discussed in Chapter 4. However, the QoS mechanisms on their own do not guarantee that contracted performance levels are delivered. That chapter highlighted the importance of understanding the impact of all parameters and their interaction. On top of that, accurate capacity planning is vital to ensure that each class meets its performance targets.

This chapter focuses on two key activities for capacity planning in which performance engineering has an important role to play. The first is the development of modelling tools that provide accurate performance prediction for a wide range of scenarios. The second is the detailed analysis of network and traffic measurements. Both are essential in achieving an optimum balance between the cost of a network and the performance experienced by the users.

9.2   Network modelling

Modelling tools that can simulate the behaviour of the entire network are invaluable for the successful design and operation of complex IP networks. By constructing appropriate views of the performance of the network under varying conditions, essential information is provided for the network operator on how to optimise resources.

Full network performance modelling is very time-consuming and detailed, resulting in a wealth of data to interpret. Therefore, for day-to-day planning and operational purposes, the conclusions of such studies are generally summarised through simple guidelines on the maximum number of allowed element utilisations which will ensure that the overall network performance targets are met. Hence, understanding when individual elements will reach their threshold is critical so that capacity can be upgraded before performance degradation occurs.

The key functionality of a network planning tool is to simulate the routing in the live network. The main input consists of a representation of the network in the form of routers, links, and sources and sinks of traffic (end-points), together with a demand matrix specifying the amount of traffic being sent between pairs of endpoints. A routing model will then determine how each traffic demand travels through the network. From this the load on each link and router can be calculated.

These simulated loads provide the user with information about which network elements approach or exceed their utilisation threshold and hence form potential bottle-necks. They can get further insight into why a hotspot occurs, by analysing which traffic demands use this particular link or router. If the traffic matrix provides per-service or per-customer information, this can then be used for a more detailed analysis. In a QoS-enabled network, maintaining performance objectives relies upon a proactive process to ensure that the utilisation thresholds per QoS class are not exceeded. Hence, in this case, per-class utilisation predictions are required.

The basic capability of calculating routes and link loads for any combination of network topology and traffic demand matrix enables designers and planners to explore an extensive range of 'what if' scenarios for the current network and future designs.

- Performance under failure conditions

  One of the most important applications is to test how the network will perform under failure conditions. Especially for QoS-enabled networks, it is of great importance to provide a reliable and consistent performance even when a network element fails or is temporarily taken out of service for maintenance. The planner needs to ensure that there is sufficient bandwidth available so that performance targets are still met.

Most network modelling tools will provide a systematic single-failure analysis, where each router and link is failed in turn. For each scenario the routing model will determine how traffic is re-routed and calculate new loads. This allows the planner to assess what the impact of any failure will be.

- Validation of network changes

A modelling tool enables the operator to test and validate changes before implementation in the live network. For an IP network it is particularly important to test the impact on routing, as small changes may often lead to unexpected and undesirable re-routings. For example, with open shortest path first (OSPF) routing, each link is assigned a weight and traffic is routed over the path with the minimum total weight. OSPF weights are often inversely proportional to link speed. Hence, if a link's OSPF weight is decreased after an upgrade, this may lead to the re-routing of some traffic flows that previously did not use this link. This re-routing could be very inefficient, leading to considerably larger delays for some flows, and, in a worst-case scenario, the upgraded link might have a higher utilisation than before!

- Traffic assessment

By applying growth factors to the traffic matrix, one can assess when the current capacity will be insufficient and plan for upgrades. It is also important to understand how sensitive the network is with respect to other changes in the traffic matrix, such as the distribution of the traffic between end-points, and, in a QoS-enabled network, the class mix. These parameters will vary for the real traffic, hence the network needs to be sufficiently robust to meet traffic fluctuations.

9.2.1   IP modelling tools

BT has developed a set of software tools with extensive modelling capabilities to enable detailed performance prediction for a wide range of IP networks. The main tools are INCEPT (IP Network Capacity Evolution Planning Tool) and IMPRESS (Internet Model of Performance and Routing of End-to-End Services). Both are highly flexible tools that can be expanded rapidly to include new technologies and services.

The key functionality of INCEPT is the construction of traffic matrices for future network scenarios. As discussed in more detail in section 9.3, predicting how the traffic on the network will evolve is a major challenge for capacity planners.

INCEPT uses current customer connections or forecasts as a basis for its traffic matrix generation algorithms, but allows the user to vary a great number of parameters so that their varying impact on network load can be explored.

INCEPT also contains a routing model which can be applied to the newly generated traffic matrix in order to predict the utilisation of all network components. This allows a first identification of potential bottle-necks.

IMPRESS builds upon INCEPT by providing a more detailed performance prediction for a range of 'what if' situations, most importantly failure scenarios. For example, the impact of any single failure can be rapidly understood by running a systematic single failure analysis.

IMPRESS is particularly aimed at QoS-enabled networks, since, in addition to utilisation statistics, it predicts end-to-end performance levels for each QoS class. To that end, analytical performance models of scheduling and buffer management mechanisms at the router output interfaces are included. These predict per-class delay, loss, and delay variation (jitter) on a link-by-link basis. The per-link performance characteristics are then combined to provide end-to-end statistics for individual traffic flows, allowing operators to validate their network against performance targets.

In addition to OSPF, the routing model includes user-selectable 'policies' which can be applied to emulate other interior and exterior gateway routing protocols such as the border gateway protocol. Furthermore, IMPRESS includes advanced features to model traffic engineering using the MPLS technology, which is discussed in more detail in section 9.2.2.2.

The tools have extensive functionality to analyse results and to answer such questions as the following.

- Which customers contribute most to congestion?

- What is the mix of services or QoS classes on a particular link?

- Which traffic flows are affected by a particular failure?

A graphical interface facilitates the analysis by colouring links and traffic flows according to user-defined characteristics, such as utilisation or loss. This allows easy identification of hotspots, inefficient routings and traffic flows that exceed performance thresholds.

9.2.2   Applications

INCEPT and IMPRESS have been used extensively to support the design and planning of BT's UK and international IP networks. Applications include the provision of capacity planning guidelines, validation of routing designs and SLGs, and studies into the benefits of MPLS traffic engineering.

9.2.2.1   SLG validation

BT's UK and global MPLS networks support multiple classes of service [1] which are backed by SLGs on delay, throughput and jitter. In order to be able to offer service levels that are both achievable and attractive for the customer, it is necessary to have a detailed understanding of the delays that traffic can experience under a wide range of conditions, including failures.

The main delay components to consider in a high-speed core network are the propagation delay (speed-of-light delay) along the path, and queueing delays at the router output interfaces. The latter are strongly dependent on the traffic load and their impact should be minimised by providing sufficient bandwidth.

In a global network, a change in the path between source and destination can increase the propagation delay significantly because of the large distances involved. In fact, such changes will usually have a much larger impact on the overall delay statistics than queueing delays, unless the network is heavily congested for a long period of time. Failures will cause temporary re-routings and their impact on the performance statistics needs to be sufficiently small. Also, as mentioned earlier, the upgrade or addition of a link can cause a permanent change of path for some traffic. All changes to the network are therefore extensively checked for their impact on routing, to ensure that reasonable delays are maintained and inefficient routings are avoided as much as possible even under failure.

Delay changes may also be caused by a change in the propagation delay of a particular link in the IP network. This can happen because of re-routings in the underlying transport network (ATM, SDH). For example, when two routers are connected via an ATM virtual circuit (VC), a failure of a switch along the path that the VC follows through the ATM network, will cause the VC to be re-routed. This will usually affect the propagation delay between the two routers.

Figure 9.1 shows the effect of re-routes at the SDH layer on delays between European core nodes in the BT MPLS network. It is important to understand their impact so that they can be properly accounted for in any service level agreement.

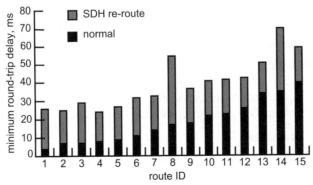

Fig 9.1   Minimum round-trip delay between European core nodes.

9.2.2.2   MPLS Traffic Engineering

In general, traffic in an IP network will always be routed along the shortest path as determined by the routing algorithm. This may lead to some links being over-utilised, while others are only lightly loaded. Ideally, in such a case one would like to distribute the traffic more evenly, to make better use of the available resources and defer investment spend for link upgrades. MPLS provides this functionality: by setting up so-called MPLS traffic engineering (TE) tunnels, the network operator can route traffic over paths other than the default ones determined by the routing protocol. An excellent overview of MPLS TE can be found in Spraggs [2].

A planning tool that can accurately simulate how traffic is routed in the presence of MPLS tunnels gives the operator valuable insight into the effect of different MPLS deployment strategies on bandwidth efficiency and performance, before implementation in the live network. IMPRESS has extensive MPLS capabilities, including static and dynamic tunnel set-up, load sharing between tunnels, and re-routing of tunnels after failure.

MPLS deployment studies were undertaken to consider the introduction of tunnels to improve load-sharing between the transatlantic links in the BT Internet backbone. At that time there was a considerable imbalance in the utilisation of the links (see Fig 9.2). By analysing a wide range of traffic growth scenarios, an optimum strategy for the deployment of MPLS TE was provided. It was shown that, because of the improvement of overall network usage, the introduction of MPLS tunnels would require significantly less capacity to support growth and, hence, reduce capital expenditure.

9.2.3   Network Optimisation

INCEPT and IMPRESS are modelling tools that allow the user to simulate how the network will perform under various operational loads. In addition, a number of design tools have been developed that optimise various parts of the network. These tools use advanced optimisation techniques such as simulated annealing and genetic algorithms. Applications include the following.

- Tunnel placement

  The optimum placement of tunnels for traffic engineering in an MPLS-based IP network ensures that the maximum utilisation of any link is minimised. Additional constraints, for example on the maximum increase in the route length compared with the OSPF shortest path, can be taken into account as well.

- Customer connections

  The optimum allocation of customer connections to access routers in the provider's network involves minimising the number of access routers needed to connect all customer sites, taking into account the planning guidelines on the

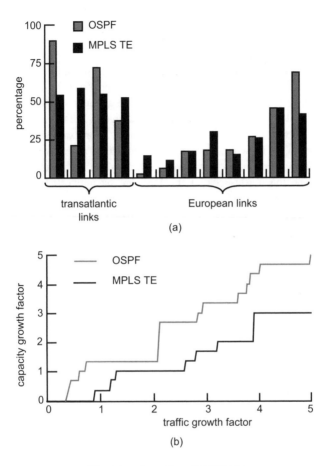

Fig 9.2 Benefits of MPLS TE.

maximum amount of customer bandwidth that can be connected to each router and the (usually distance-related) cost of the customer connections.

9.3 Traffic forecasting

In order to plan any network successfully, one needs to be able to predict accurately when and where bottle-necks will occur, so that capacity can be added in time and in the right place. It is therefore vital to understand how the network is currently used, and how it is likely to evolve.

The network needs to be dimensioned so that the maximum load of a normal traffic profile can be carried without performance degradation. It falls outside the scope of this chapter to discuss in detail how a 'normal' traffic profile should be defined, but one important aspect is the time-scale over which traffic is considered. As shown in Fig 2.11 in Chapter 2, profiles over longer time-scales will exhibit lower peaks because of time-averaging.

One also has to decide when a peak is considered 'normal'. For example, data over a longer period would be needed to see if the peak traffic load in Fig 9.3 is a regular occurrence and hence should be taken into account when dimensioning the network. Obviously, it is not economical to dimension the network for extremely high traffic loads that may occur only occasionally. For example, an IP backbone may see a surge in traffic when many users try to access a few Web sites in the case of a disaster or high-profile sporting event. Although in such a case the servers are most likely to be the bottle-neck, if the load on the network becomes too large, routers will start to drop traffic. This results in users experiencing extremely slow downloads and possibly time-outs, which (hopefully) encourages them to disconnect. In a QoS-enabled network, the QoS mechanisms (policing and priority scheduling — see Chapter 4) will ensure that higher priority traffic still experiences an acceptable performance.

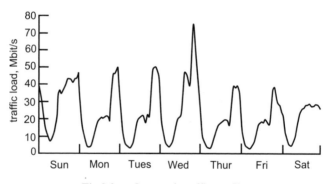

Fig 9.3   One week traffic profile.

In order to get an insight into the overall network traffic profile, a carrier-scale IP network operator needs to understand the particular characteristics of each service.

- What is the maximum amount of traffic offered at each core node?

- When do the peaks occur?

- What is the distribution matrix, i.e. where does the traffic go?

- For a QoS-enabled network, what is the class mix?

- What are the growth trends?

For example, it is important to know whether the peak periods for different services or different groups of customers coincide as this will affect the maximum amount of traffic that is likely to be offered at any one time and hence how the network should be dimensioned.

Figure 9.4 shows typical Internet traffic profiles for business and residential users during weekdays. Unsurprisingly, the busy period for business traffic is between 9 a.m. and 5 p.m., whereas residential traffic peaks in the evening. If capacity planning were based on the assumption that business and residential traffic peak at the same time, then for the (unit-less) traffic profiles in Fig 9.4 the network would be dimensioned for a maximum load of 5.5. This would overestimate the actual peak of 3.9 by more than 40%!

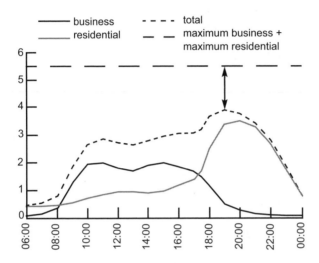

Fig 9.4   Daily profiles for business and residential Internet traffic.

For an existing service, modelling of future scenarios will be based primarily on the current traffic profile and predicted growth. Chapter 10 shows (in Fig 10.2) how to combine forecasts and historic traffic data to predict when the current capacity will be insufficient. However, it is important to consider any differences between the current and modelled scenario and assess how they are likely to affect the traffic profile. For example, changes to the network topology, such as the addition of extra nodes, will affect the traffic distribution. Also, growth trends are likely to be affected by a price change or by the introduction of a new service.

One of the most challenging aspects of traffic forecasting is predicting what the likely distribution of traffic across the network will be, especially for a new network or service for which no historic data is available.

For example, in an Internet backbone most of the traffic flows will be to and from peering points, where traffic is exchanged with other network providers. Since any network, however large, will only form a small part of the global Internet, most of the traffic from backbone customers will have a destination on another network, hence will leave at a peering point. How the traffic is split between the different peering points will therefore have a major impact on the utilisation of the core network elements.

Since it is impossible to know exactly where the traffic will go, the network should be planned in such a way that it is sufficiently robust against fluctuations in the traffic pattern. The INCEPT tool enables a rapid turnaround of scenarios with different assumptions on traffic distributions to assist in the planning. For each service, the user can vary a number of parameters, such as the ratio between on- and off-net traffic, the percentage of off-net traffic sent to a particular peering point, and the ratio between upstream and downstream traffic.

Planners of IP VPN platforms face even more complex challenges when building traffic matrices. For example, BT's IP VPN services are mostly any-to-any services. This means that the customer only has to specify the maximum amount of traffic each site will send into the network, not how much traffic will be sent between different sites. This provides the customer with much greater flexibility than a point-to-point service such as used in ATM and frame relay. For those services, virtual circuits (VCs) are set up between each pair of sites, which require the customer to specify the mean and maximum amount of offered traffic on a per-VC basis.

It is important, especially for a QoS-enabled IP VPN network, to understand the effects of the addition of a new (large) customer on the overall performance of the network. To investigate this, assumptions will need to be made about the extra traffic that the customer will add to the network, in particular how much traffic will be sent between the different sites. The two main connectivity profiles for an IP VPN are hub-and-spoke and mesh. In a hub-and-spoke configuration all traffic is sent to and from one or a small number of main sites. This profile applies to many types of business with a main office or data centre and many local branches, e.g. a banking organisation. A completely different profile arises in a mesh configuration where all customer sites send traffic to each other. To complicate things further, a customer might have different traffic profiles for different applications. For example, most of the data traffic might be to and from a data centre, whereas voice traffic might be sent between any pair of sites.

INCEPT provides a suite of algorithms that generate individual traffic matrices for each customer. These use a number of parameters such as access speeds to predict the traffic distribution. The underlying assumptions are regularly validated against real traffic measurements.

In general, it is important to validate assumptions for predictive modelling by comparing modelled against measured utilisation figures and traffic patterns for the current network. This needs to be a regular process as traffic profiles are likely to change over time.

9.4   Measurements

Traffic and performance monitoring is crucial to the management of any IP network. Firstly, (near) real-time monitoring is required so that any major problems such as router or link failures are quickly discovered. In addition, there should be an ongoing collection of measurements to support the planning process. This gives vital information about current performance, such as link and router processor utilisation, traffic profiles, and trends. Suitable interpretation of the statistics also provides more detailed understanding of what affects the performance of the network, which should be used to improve design and planning guidelines.

There are two types of measurement — active and passive. In an active measurement system, probe packets are injected into the network to measure delay and loss along specific paths. The best-known example of this is 'ping'. More sophisticated measurements are, for example, provided by Cisco's Service Assurance Agent (SAA) [3]. SAA jitter probes give an extensive set of both one-way and round-trip delay, loss, and jitter statistics for the path between the source and destination router. In a QoS-enabled network, measurements per QoS class are essential to understand per-class performance, and ensure that performance targets are met.

Many Internet backbone operators already provide more or less detailed information about the performance of their core network based on probe measurements. For example, BT publishes hourly loss and delay statistics between core nodes in the Internet backbone [4].

Active measurements can only provide a partial view of the performance of the network and should be complemented by performance data from individual network elements. For example, routers keep track of an extensive set of performance data, such as availability, link and router processor utilisation, the number of dropped packets, etc. This information can be retrieved by a management station, which will poll each device at regular intervals. Systems such as Concord eHealth [5] can manage large numbers of network elements and enable the user to run reports on the collected data to extract a wide variety of information, such as trends, top-N lists, at-a-glance reports of current network performance, or detailed performance statistics per network element.

In addition, traffic flow data is invaluable to get a detailed understanding of the traffic profile. For example, the NetFlow feature in Cisco routers collects flow-level records on individual interfaces [3]. The records contain information about the traffic end-points, next-hop information, type-of-service bits, and number of bytes and packets per flow. Although by no means trivial, these can be converted into a core-to-core traffic matrix detailing the traffic volumes between each pair of core nodes. In addition, the data can be used for more detailed traffic characterisation such as a breakdown per QoS class or per application. NetFlow can generate large amounts of data and hence a careful design for collection and aggregation of flow information is needed to avoid a degradation of the overall performance of the network.

9.4.1 Measurement Analysis

On BT's IP networks a significant amount of routine measurement takes place on a continuous basis. Network devices are usually polled about once every 5 minutes.

Probes are sent into the network with a frequency of at least once a minute. It is important that the probe frequency is high enough so that a sufficiently detailed view of the performance of the network is obtained. On the other hand, the extra traffic caused by the measurements should be sufficiently small so that customer traffic is not affected. A similar trade-off has to be made when deciding how often routers and switches should be polled.

Platform managers will receive performance reports in which the data is averaged over much longer periods, such as a week. It is usually the task of the network performance engineer to undertake detailed analysis of the low-level data in order to provide insight into how performance may fluctuate over short time-scales and flag up any unexpected behaviour. This requires a great deal of expertise in network and traffic behaviour in order to be able to recognise what can be considered as normal and what needs further investigation. For example, detailed analysis of delay measurements can reveal incorrect routing configurations or the occurrence of lower-layer re-routes, as shown by Fig 2.9 of Chapter 2. In the same chapter, Fig 2.10 depicts short periods during which the performance of real-time applications such as voice will be severely degraded because of extremely high jitter values.

Almost inevitably, measurement data will contain some anomalies, often caused by the measurement process itself. For example, performance variables in the router may occasionally give spurious values. These could have a significant impact on longer-term averages, resulting in daily or weekly reports wrongly suggesting a performance degradation of the network. Because of this, it is important that data is screened at a fine granularity in order to establish whether spurious values have occured and assess their impact on the overall performance statistics.

Network measurements are also essential for the validation and refinement of design and planning guidelines. As discussed in Chapter 4, these will initially be based on the outcome of laboratory testing, analytical models, or simulation. Inevitably, in all cases considerable simplification with respect to the real network and traffic characteristics will have been made, so it is vital that the predicted performance is compared with what actually happens on the network. Any significant difference should lead to a re-assessment of the planning rules. Furthermore, it is important to investigate what caused the difference, so that limitations of the test environment or performance models can be better accounted for in future recommendations.

9.4.2   Case Study — Queueing Delays in an Internet Backbone

In order to maximise network resources, it is of great importance to understand at what utilisation levels the network can be run without significant performance degradation. At higher loads, there is an increased probability that packets have to be buffered before being transmitted and hence will experience additional delays.

The relation between utilisation and queueing delays is strongly dependent on the burstiness of the traffic, i.e. the peak-to-mean ratio of packet arrivals at the output buffer. In a core network, this can be affected by a great number of factors such as the mix of applications, user behaviour, the degree of multiplexing and the degree of contention experienced on the access links, but their precise impact is not well understood. In section 2.6 of Chapter 2, other complexities associated with IP traffic modelling are pointed out, in particular the adaptive behaviour of the TCP protocol, which is used for the majority of traffic streams in an IP network. Expert analysis of data from the live network can greatly improve understanding of the traffic characteristics, and is therefore invaluable for the development of robust operational performance models.

Figure 9.5 is the result of a detailed analysis of delay and utilisation measurements on a T3 (45 Mbit/s) core link in an Internet backbone. During several days the round-trip delay on the link was measured at a frequency of once per second and a resolution of 0.1 ms. At the same time the most heavily utilised of the two interfaces was polled once per second for the number of bytes transmitted out of the interface. Although in theory this would give average utilisation levels over one second, in practice it appeared that the smallest time-scale over which a meaningful value for the average utilisation could be obtained was approximately 10 sec. This was attributed to the distributed architecture of the router, where data is collected on the processors on the cards and, with relatively low priority, transferred to the central processor. The link reached extremely high loads during the measurement period (a rare opportunity for a performance engineer!) and was upgraded soon afterwards.

The analysis concentrated on delay statistics for one-minute intervals. For a 45 Mbit/s link, considerable fluctuation of the traffic arrival statistics, and hence of

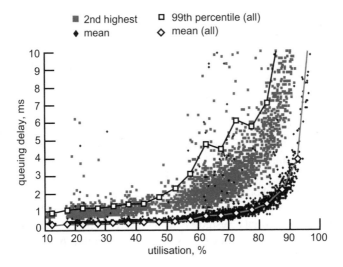

Fig 9.5   Utilisation versus queueing delay.

the queueing delays, can occur on a sub-second time-scale. However, the main objective here was to provide input into design and planning guidelines, which consider average utilisation levels measured over significantly longer periods.

In the two scatter graphs in Fig 9.5 each data point represents a one-minute interval, with average utilisation plotted against mean and second highest queueing delay over the 60 delay measurements. The second highest was chosen instead of the maximum to remove the impact of some unexpectedly high and possibly erroneous values. The line graphs are the result of further aggregation, where the mean and 99th percentile were calculated over a number of measurements ranging from about 1600 for 15% utilisation to more than 24 000 for 70% utilisation. These graphs provide a very useful model for future performance prediction.

9.4.3   IP QoS

Detailed analysis of measurements as in the above example are vital in providing QoS in a cost-effective way. Since IP QoS is a new technology and little experience has yet been gained with such networks, planning guidelines will initially be conservative to ensure that the performance targets are readily met. However, it is expected that with experience and with better understanding of traffic profiles and network performance, these guidelines can be relaxed so that network resources can be used more efficiently. Close monitoring of the network is therefore required in order to gain a detailed understanding of how performance is affected by different factors. Furthermore, this will need to be an ongoing process as the network and traffic characteristics evolve over time.

9.5   Summary

With the growing performance requirements on carrier and corporate IP networks, operators need an increasingly more accurate understanding of how their network performs, what the traffic characteristics are and how these are likely to evolve, in order to successfully manage the network. This chapter has discussed the importance of modelling and measurement analysis in the planning process. Both are areas where the skill and expertise of the performance engineer can make a significant contribution.

Network modelling tools have already been successfully used in the design and planning of BT's IP networks, reducing investment spend and increasing confidence that the required performance targets will be met. Detailed analysis of measurements has given vital insights into the performance of the network, which have led to refined planning guidelines and more accurate performance models. Moreover, with the growth of IP QoS, both activities are expected to play an increasingly important role.

References

1   BT's Global Services — http://www.btglobalservices.com/en/products/

2   Spraggs, S.: '*Traffic engineering*', in Willis, P. J. (Ed): '*Carrier-scale IP networks: designing and operating Internet networks*', The Institution of Electrical Engineers, London, pp 235-260 (2001).

3   Cisco — http://www.cisco.com

4   BT's Internet Backbone Performance Measures — http://ippm.ignite.net/

5   Concord eHealth — http://www.concord.com

10

PERFORMANCE MONITORING FOR IN-LIFE CAPACITY MANAGEMENT

D J Chown and J Graham

10.1 Introduction

Communications services are becoming increasingly more complex and geographically dispersed. Local administrators and managers have only a limited view of the interactions between the many different platforms, subsystems and software. However, overall service to the end customer obviously needs all the components to perform and interwork smoothly. Therefore, all such systems need a responsible co-ordinating team to oversee the total technology and business machine, and to maintain a detailed check on the performance of the whole network as experienced by its users. Providing this co-ordinating team with a management view of performance, and all it entails, is difficult. Some of the issues and solutions, evolved over a number of years, are discussed below.

The graphs and data are for illustrative purposes only.

10.1.1 Product Life Cycle

The importance of specifying performance requirements and including performance engineering techniques early in the life cycle of a product, that is, during the feasibility and design stages, is increasingly accepted and understood. It is also accepted practice to devote resources to performance engineering throughout the development and testing phases. However, less attention is generally given to monitoring, reporting, performance management and capacity planning during the 'in-life' phase of the product. Application of these techniques provides an immediate, highly visible benefit in that they enable severe problems to be corrected as soon as they occur, and many potential problems to be identified and resolved before they actually occur. This requires information, but it is not sufficient simply to make it available — the information has to be presented. More emphatically, the right information has to be presented to the right people at the right time. This can be achieved and the benefits can be enormous.

For new systems, it is essential to know the capacity of components beforehand. This knowledge can be acquired by measurement under controlled conditions, or by modelling on a computer, before deployment. Only in controlled experiments, with realistic prototypes or models, can engineers ensure the implementation is optimal.

In-life systems provide only limited access for performance engineering work — engineers cannot make arbitrary experimental adjustments nor load the systems with generated traffic, and any results they do get are subject to the idiosyncrasies of the particular installation. Therefore, the results cannot readily be applied to all such systems.

A system that has been properly measured by performance experts can be dimensioned more accurately, and the results provide a basis for questioning in-life performance.

The aim of including performance engineering at all stages of the product life cycle is to forestall problems. This applies equally to in-life systems.

10.1.2   Preparing for Growth — or Shrinkage

Successful services tend to grow, whereas unsuccessful ones usually shrink to nothing. By monitoring and forecasting growth at the same time as knowing about capacity, managers have early warning of the need for system enhancements. This can be very valuable because long lead-times sometimes apply when ordering specialised equipment.

However, in complex systems, some parts can grow while others shrink. As technology advances, it becomes expedient to introduce new components, in parallel with the old, performing essentially the same functions. They may be cheaper to run and maintain, faster, bigger or they may enable service enhancements. Eventually, the old components may be 'phased out' by a scheduled migration of the workload to the new systems, but scheduling the closure requires planning, monitoring and reporting.

Figure 10.1 shows an example of the introduction of a new transaction type, denoted 'Type B'. Firstly the new system picks up the load from the old, and, shortly after that, transaction Type B is introduced.

10.1.3   Fault Detection

Fault detection is a useful by-product of performance engineering. This is frequently demonstrated in performance testing, because performance testing is only feasible

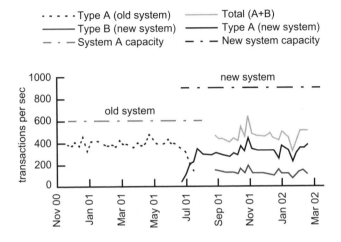

Fig 10.1 Introduction of a new system to process a new transaction type.

on working systems. However, facilities, such as TeleMarketing Services, are designed to be fault-tolerant, which means faults usually manifest themselves as degraded services. Fault-tolerant systems are a special case in performance measurement, because of the need to determine the capacity, and specify the performance requirements, under specific fault conditions.

10.1.4   Positive Reporting

There are several benefits from reporting both bad and good news:

- when systems require remedial engineering intervention, engineers and managers can evaluate the usefulness of their work;

- senior customer-facing staff have immediate access to pertinent, up-to-date information;

- by receiving good results as well as bad, managers enjoy the satisfaction of knowing their systems are working well and that the monitoring process is working — it will warn them when their systems are working less well.

There is also a less obvious benefit. The results from all facets of the business and technology system are visible to all those managers closely connected with the project. Most managers welcome the opportunity to discuss issues with others who have parallel interests and common objectives.

In summary, reporting bad results has many benefits, while reporting good results dispels doubt and reinforces good practice.

10.1.5 Consistent Format

To realise the above benefits, it is not enough simply to put the results on a Web site for component managers to browse. The results must be presented, and the presentation needs to be sharp and direct, requiring a minimum of managers' time and preparation. The production of such a report is the last stage of the collecting, sifting, condensing and summarising that takes place before the information finds itself in the managers' (metaphorical) in-trays.

The report needs a memorable name to facilitate and encourage discussion, but, for the purpose of this chapter, it is called the 'Capacity Report'.

The graphs and tables need to be updated periodically, typically, every month. The component managers themselves ultimately decide what should appear in the Capacity Report. The emphasis is on simplicity, which is essential for rapid assimilation. Readers quickly learn to identify the graphs and recognise potential problems.

Naturally, with an evolving, complex system, there are frequent changes, but the layout needs to remain essentially the same. This is necessary for production of the report, as it requires a high degree of automation to complete the work within an acceptable reporting period.

Figure 10.2 is a graph adapted from a real example. It shows the calling rates of two types of call. Type A could be, say, simple calls to service providers with no special routing facilities, whereas Type B could be calls to service providers with several answering centres requiring further resources to route the calls.

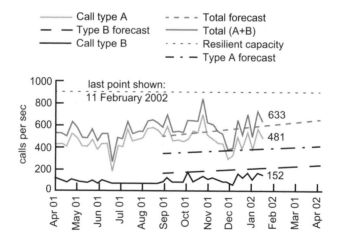

Fig 10.2 A typical graph showing measured calling rates, forecasts and system capacity.

In Fig 10.2, the calling rates are growing approximately in line with the forecasts, but Type B calls have failed to reach the calling rate forecast. The resilient capacity is not threatened in the near future. The parameters on the graph would be shown as 'green' in the 'RAG table' (RAG stands for 'red-amber-green' — see section 10.3.3 for more details).

10.1.6   Timing

Assuming reporting periods of one month, the time from the end of any month to completing the Capacity Report cannot exceed four weeks. In practice, there are usually about 15 working days in which to gather and process the data, and write and present the results.

As far as gathering the data is concerned, the first few days of the month are a slow time, because various parts of the organisation have to contribute. It takes time to get the source data, possibly do some preliminary processing and to 'make it available' to the author of the Capacity Report. For example, the highly automated system for collecting call samples from all the BT exchanges in the country takes three to four working days. This centrally maintained database is one of the major sources of data for the Capacity Report.

The time between presenting this report and starting the next is not a 'dead time'. It is a time to follow up unexpected results and seek explanations, or carry out investigative analyses. It is also a time to enhance and streamline the data-processing procedures for producing the Capacity Report.

10.2   TeleMarketing Services

The remaining sections describe the performance mechanisms used within the network for monitoring, reporting and controlling BT TeleMarketing Services. These services provide a very clear example of the use of the above generic performance principles and techniques.

TeleMarketing Services is a telephone facility with various features and payment structures. It encompasses all BT non-geographic number ranges and their associated network services, such as Freefone, NationalCall and ValueCall. A 'number range' is simply a set of telephone numbers with the same initial digits, such as 0800, 0845 and 0870. These particular number ranges are not tied to any specific locality and are therefore termed 'non-geographic'.

Organisations choose these special numbers to make it easier to provide services and information to their customers and other callers.

TeleMarketing Services provides the convenience of a single dialled number for callers, while the network routing products and services provide advanced features for called customers, that is, the service providers. Those service providers who choose advanced features, have an advanced features call plan (or simply, 'call plan'), which enables them to control how calls are distributed among their answering centres. Calls can be distributed according to time-of-day, by proportion, etc. To change how calls are handled, service providers can use a computer link, or an interactive voice-activated system called Select Link.

The TeleMarketing Services advanced features give organisations flexibility and versatility in the way they manage their TeleMarketing calls. This means they can answer more calls more effectively, capturing more business while keeping costs to a minimum.

10.2.1   The Complexity of TeleMarketing Services

The whole TeleMarketing Services structure is very large and diverse, and requires sophisticated, automated management and control. When the customer service, order management and billing systems are included with the technology systems, there are over 200 different types of component. However, the underlying mechanisms, greatly simplified, are illustrated in Figs 10.3 and 10.4.

The Intelligent Network (IN) is central to TeleMarketing Services. Figure 10.3 shows where the IN fits into the call routing strategy, and Fig 10.4 shows the principal management and control systems.

The IN performs the number translations that route calls to the service providers. Calls to service providers with more than one call centre are translated by a separate subsystem within the IN because the translation depends on the call plan.

In Fig 10.3, for example, a caller on a digital local exchange (DLE) dials a Freefone number, 0800... . The DLE is connected via a digital main switching unit (DMSU) to the IN. The IN contains all the 'intelligence' required to translate the Freefone number into a regular UK regional number. In this example, it is an Ipswich number, beginning 01473... , so the call is routed to the service provider through, in this case, a different DLE.

Similar routing applies to all the other areas of TeleMarketing Services.

There are several places across the UK where the IN facilities are implemented. They all have advanced routing and number translation capabilities. Calls can also be controlled by 'voice interaction'.

One of the principal benefits of the IN, apart from advanced call routing, is the facility for service providers to monitor and control the effectiveness of their call plans. It is the 'provisioning interface' that provides the necessary access to the IN, via the public switched telephone network (PSTN) (see Fig 10.4).

The provisioning interface also provides access for support staff, who can implement new orders and updates for TeleMarketing Services customers.

For the Capacity Report, the call statistics interface is essential, but its main purpose is to provide information for customers' reports. Customers may update their call plans in the light of these reports.

Many other systems also have interfaces to the IN. Perhaps the most important is BT's very substantial billing system.

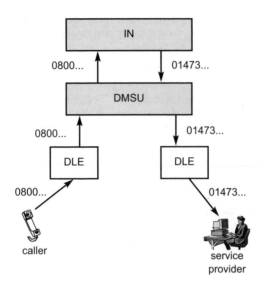

Fig 10.3   How the IN routes TeleMarketing Services calls.

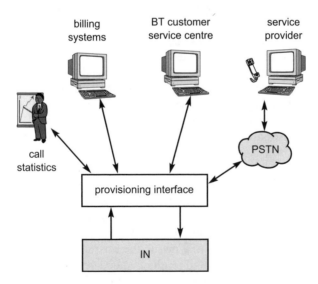

Fig 10.4   IN management and control systems.

Staff at network operations centres monitor the operation of the IN continuously. They can detect congestion and faults as soon as they occur, and can sometimes re-route traffic to avoid trouble spots. They can also react to transient overload, which can occur, for example, when a particular number is announced on television. However, they appreciate being forewarned of major phone-in events so that appropriate action can be taken in advance.

All these systems work together to provide a flexible and powerful call-handling platform that maximises the throughput of calls to the service providers, and keeps them informed by means of statistical reports. The statistics collection also provides invaluable technical information for managing the IN and associated systems.

10.3 Reporting Principles

The processes in place for TeleMarketing Services, on which this chapter is based, provide monthly updates on well over 100 parameters on about 70 graphs and tables, using over twenty different data sources. The reliable supply of good quality data is essential, and contingency plans are in place to cover all foreseeable difficulties.

10.3.1 The Data-Collection and Reporting Process

The primary purpose of the Capacity Report is to keep managers informed across a wide range of interacting and mutually dependent systems. It begins with data collection at the many sites, transmission to a central point and analysis. The detailed mechanisms are negotiated independently for each data source. Consequently, the data arrives by many different methods, but they are all in a form that can be processed immediately on a PC or under Unix. There is no manual entry of source data. The co-operation of the data providers is ensured by the expectation that useful information will appear in the report.

The recipient uses a variety of automation techniques to transform the data into the simple graphs and tables, which form the basis of the report.

As the Capacity Report gives a high-level view of capacity and performance, it does not cover the fine details of all the systems. Nor does it give the fine details of the interactions between the remote systems. The content is selected and organised carefully, and presented in a way that enables busy, time-conscious managers to absorb the information as quickly as possible. A consistent format is essential.

Text is kept to a minimum. This means there is no room for 'colouring' the results. If results are bad, it says so. If results are good, it says so. If the data failed to arrive, it says so, for all readers to see. There are no euphemisms.

The coverage of the report is ultimately determined by discussion in the Capacity Forum. (The Capacity Forum is a monthly meeting of the co-ordinating team that oversees the systems and services. It has a chairman, representatives from major facets of the business providing the TeleMarketing Services, and a presenter for the Capacity Report. Usually, invited consultants or advisors are also present.) The forum chooses parameters to reflect the needs of the particular component or facility manager. In general terms, these are aimed at:

- forecasting capacity exhaustion and scheduling system enhancements;
- detecting degradation of service;
- monitoring recurring problems;
- monitoring workload migration to newly installed systems;
- monitoring workload sharing and balancing;
- monitoring the quality of service provided to major customers;
- monitoring the impact of engineering work, such as fault clearance;
- dimensioning of future enhancements.

10.3.2   Readership of the Capacity Report

Although the Capacity Report is produced and presented by its compilers, its readership is controlled by the chairman of the Capacity Forum. The information it contains is aimed specifically at senior managers or platform managers, and not the 'shop floor'. On its own, the Capacity Report is insufficient for day-to-day system administration, but when there are problems affecting the business, it alerts managers to the need for further investigation and possible remedial action. At the Capacity Forum, a co-ordinated response can be agreed across the various platforms.

The entries that appear in the Capacity Report are under constant review. The Capacity Forum decides on changes, but sometimes changes are 'experimental'. If a hitherto unmonitored feature becomes important, through the addition of new services, for example, the Capacity Forum may need several new graphs in the report. Then, over the subsequent few months, the graphs will be changed or removed to leave those that are the most informative.

This transitional period can be difficult, because there is no automation in place for processing the data, but the ultimate aim is to arrive at a simple informative graph of a few parameters. As an extreme example, system-level measures, like CPU occupancy and disc-writes, may be rejected in favour of user-oriented parameters like 'response time' and 'throughput' or high-level parameters, like 'concurrent customers on-line' and 'new orders processed'.

10.3.3   RAG Table

Produced in colour, the RAG table gives an 'eye-catching' overall view. It gives a 'rough and ready' picture of each measured parameter over the past 13 months:

- red, meaning 'immediate attention required';
- amber, meaning 'some action may be required soon';
- green, meaning 'no apparent problems'.

There is also an 'Alert' entry in the table to draw attention to any other unexpected behaviour.

As with every section of the Capacity Report, the production of the RAG table is automated as far as possible without compromising its adaptability.

When new sections are added to the Capacity Report, new entries are also added to the RAG table. Each parameter to appear in the table has thresholds set to define the red-amber-green criteria. Neither the thresholds nor the working methods are shown in the report, but they may be discussed at the Capacity Forum. Figure 10.5 is an extract from the RAG table. The complete table has over 100 entries.

Red-amber-green capacity and volume summary table																	
N	D	J	F	M	A	M	J	J	A	S	O	N	Alert	Section	Section title	Statistic
														Part 1		
G	G	G	G	G	G	G	G	G	G	G	G	G		2.1.1	Weekly call volumes	TeleMarketing
G	G	G	G	G	G	A	G	A	A	A	A	G		2.1.1	Weekly call volumes	Internet calls
G	G	G	G	G	G	G	G	G	A	A	G	G		2.1.2	Weekly Lo-call volumes	BT 0845 range
G	G	G	A	G	A	A	A	A	G	G	G	G	Alert	2.2.1	Monthly Freefone GoS	BT 0800 numbers
														Part 2		
G	G	G	G	A	R	R	A	G	G	G	G	G		3.2.3.1 b	Platform load balancing	Calls per second
														3.7.2	Impact of switch migration	

Fig 10.5   An extract from the RAG table.

'Alerts' in the table can be used to draw attention to particular results. The criteria may be laid down beforehand, or alerts may be used in any appropriate way. They can relate to performance, spurious results, or anything requiring closer than usual attention.

10.3.4   Simple Graphs

In practice, there is an arbitrary split between network-wide parameters, such as 'telephone call volumes' and 'congestion statistics', and platform-specific parameters, such as 'simultaneous ports used' and 'database size'. These different aspects are reported in separate volumes of the Capacity Report. Not all readers require both volumes.

The platform-specific parameters are shown, wherever possible, as simple graphs. For new graphs, some technical advice from the platform managers is needed before a format for presentation can be proposed. The Capacity Forum may decide to modify the format, but, ultimately, the graphs are mainly for the specific platform managers. Therefore, there may be separate graphs to satisfy the needs of both the forum and the platform managers.

With these clear, simple graphs, managers quickly and easily form a projected visual concept of the workload, capacity and useful life of their systems.

Figure 10.6 is based on a genuine case. Note that the left-hand axis is inverted because the readers prefer to see the graph rise as demand increases. The frequent changes in direction are the result of engineering intervention to recover used space. The trend line is used to predict the would-be exhaustion date if no recovery work were undertaken. The text on the graph is generated automatically.

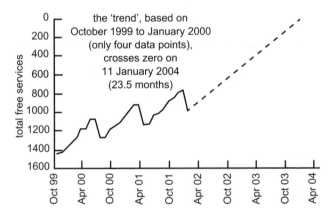

Fig 10.6   An example showing the benefit of intervention by engineers.

10.3.5   *Ad Hoc* Graphs

A few graphs appear in the Capacity Report to deal with transient situations.

For example, an ordering system was running out-of-order numbers because the total possible number length was insufficient to meet the forecast number of sales. It would take time to modify the computer programs and all the associated business systems to cope with the change. In the meantime, a graph appeared in the Capacity Report to show the projected exhaustion date. If the new systems had not been in place in time, it would have been very costly, but at least managers would have been forewarned.

10.3.6   What is Monitored

The Capacity Report covers parameters that range from the 'obvious' to the 'esoteric' in order to meet the varied needs of the platform managers, but there is a strong emphasis on end-to-end performance.

Some obvious parameters are call volumes, calling rates, grades of service, ineffective calls, traffic, holding times, circuit occupancy, distribution of circuit occupancy, proportion of call types, stored translations, closeness of balance, message throughput, data volumes, 'data ready' times, orders received, concurrent users, service availability.

10.4   Presentation Techniques

The aim is to provide easily digested, useful information. Nearly all the results are shown as graphs, but there are also a number of tables and some background information.

Each graph or table is preceded by a brief description of what it shows. There may be some platform information, but this is very brief. Platform managers know much more about their respective areas than the author of the Capacity Report. However, the author knows how to process the data and has knowledge and experience of statistics. The author also has access to expertise in all the relevant technologies, including data management, computing and communications. Therefore, the Capacity Report states the source of the data, and gives a high-level description of how the data has been used, for example (illustrative only):

> Source: sampled calls database. The data does not contain information about the caller.

> The graph shows the proportion of calls that failed due to congestion between 10 a.m. and 11 a.m. on the busiest Monday of the month.

Specialist statistical terms are avoided, not because the readers may not understand them, but because it is a worthy discipline to explain the concepts in plain English. Statistical terms are easily misunderstood, and avoiding jargon helps to ensure that all readers draw the correct conclusions from the analysis. It also gives confidence that the data is being analysed correctly.

Following each graph is a short description or explanation of the latest results. If there is an alert in the RAG table (see section 10.3.3), then a reason is given here. This is also the place for explanations if the data is unavailable for any reason.

10.4.1   Tracking Major Customers

Some dialled numbers regularly receive very large numbers of calls, but, in addition, special events can generate a huge intensity of calls to a single dialled number. The impact of these events is monitored carefully, and the results are reflected in the Capacity Report. Calls to these numbers can affect the service, for short periods, but there are sophisticated mechanisms in place to protect the network, and other users, from very large surges of traffic.

The call statistics make it possible to pick out called numbers that receive very high volumes of traffic, and examine the quality of service they experience. Some high-usage numbers are also identified as being representative of high-usage customers in general. They are known as the 'top twenty' in each number range. They are chosen firstly based on how many calls they receive, and then agreed in the Capacity Forum. Once agreed, the list remains essentially unchanged from month to month.

10.4.2   Detecting Resource Wastage

It is not unknown for service providers to publish incorrect numbers in their advertisements. This has sometimes led to huge numbers of calls to unallocated telephone numbers, or other inappropriate numbers. Such calls 'waste resource' that would otherwise be available to revenue-earning services.

To show this in the Capacity Report, there is a table that contains the dialled numbers that resulted in the most ineffective calls. In fact, there are two tables — one with the highest **percentage** of ineffective calls and one with the highest **number** of ineffective calls. In order to keep track of numbers that continue to appear, their position in the previous month's table is also recorded.

Overprovisioning is another way in which resource wastage can occur, but this is a different concept where equipment is working well below its capacity and may be idle much of the time. Overprovisioning is something to be avoided rather than corrected. Sometimes surplus capacity can be assigned to other uses, but it is rarely cost-effective to remove it unless it incurs other costs such as excessive rent or maintenance.

The Capacity Report sets out to expose over-provisioning in certain areas. As an example, special trunk routes link exchanges that use older technology to new exchanges, while the older exchanges are progressively decommissioned. To optimise costs and throughput, the occupancy of these routes needs to be kept within an acceptable range. The distribution of route occupancy in the report provides the information required.

10.4.3   Load Balancing

Load balancing or sharing is an important concept throughout the TeleMarketing Services network, and, of course, in performance engineering — two explicit examples are given below.

- Number translation

   TeleMarketing Services calls have to be routed to the service providers. Briefly, this means the dialled numbers have to be translated to routing digits. There are various sites around the country sharing this load. The Capacity Report compares the load at these sites.

Incidentally, balancing the workload of these sites is important for another reason. They are designed for very high resilience, such that if one site is out of service, the remaining sites can carry all the workload. If imbalance were tolerated, the ability of the remaining sites to handle the total load would be reduced.

- Signalling links

  The second example involves a set of signalling links. Balancing was required for the usual reason — to make best possible use of the available resource. It happened that the Capacity Report showed a severe imbalance. When engineers investigated, they found a configuration error in the terminating equipment. They corrected the error and took steps to prevent a recurrence. Figure 10.7 is derived from the graph that revealed the problem.

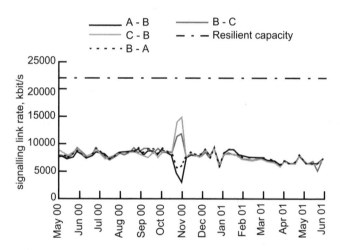

Fig 10.7   Workload imbalance reveals a configuration error in terminating equipment.

10.5   Business Benefits

As the Capacity Report does not earn revenue directly, its 'business' benefits arise from its 'technical' and 'managerial' benefits. It is a technical report directed at managers.

The Capacity Report has greatly assisted in performance management and in-life development, while, on the management side, it has facilitated the scheduling of major changes, enabled the co-ordination of management strategy and kept senior customer-facing staff informed of network performance issues.

10.5.1   Dedicated Reporting

The idea of the Capacity Report is a simple one. The statistical reporting is centralised for efficiency. It is nearly always more straightforward for individual component managers to supply raw data, than to produce formal reports. The Capacity Report also highlights the most important and informative parameters that might otherwise be obscured by a myriad of local issues.

So, why not concentrate the effort required for all the components into one dedicated work area? That way, a very small number of people could develop the necessary skills and apply them, as far as possible, across all the platforms. There has to be a period of learning for each component, but thereafter, local staff need only provide the data. Very often, the provision of data can be automated. It could be sent by an overnight, automated file transfer, for example. System administrators usually welcome the opportunity to offload the reporting task and concentrate on their regular work, while having confidence that the data is in the hands of someone who specialises in statistical reporting. Overall, this is a very economical way of maintaining the momentum of the capacity review process.

10.5.2   Dimensioning

The experience of the Capacity Report enables new additions to be dimensioned as the systems grow, but the new additions may employ more advanced technology, such as faster processors (see Fig 10.1, for example). This does not invalidate the previous results, but it does mean that some more detailed studies may be required. The saying, 'performance does not scale up', means that there is no simple relationship between performance of systems of different size or speed. However, the Capacity Report provides a starting point, and gives a 'sanity check' on any more detailed results.

Each time an upgrade takes place, there are certain unavoidable costs, in addition to the actual hardware. The idea of accurate dimensioning is to find the optimum between a few large upgrades and many small upgrades. Obviously, over-dimensioning requires excessive capital investment and under-dimensioning leads, too soon, to further costly upgrades.

10.5.3   Scheduling Deployment

The Capacity Report is more valuable with regard to scheduling new additions. This is one of the primary aims of the report. It announces the 'capacity exhaustion date' clearly for relevant components. It says it graphically, and, as the date approaches, it says it in words. The report begins to be more emphatic when the exhaustion date is less than six months away.

Early deployment may feel secure, but it requires early capital investment with resultant loss of liquidity. Late deployment means degraded services, customer discontent, loss of revenue, and possibly, loss of customers.

10.5.4   Managing Service Enhancements

In practice, deploying new systems involves more than simply connecting up and switching on. It is rare that a new system handles only completely new growth. At the very least, arrangements have to be made to divert some or all of the workload from existing components. Figure 10.1 shows a rather rapid transfer of workload. For major technology enhancements, the changeover can take years as new devices are installed and workload is transferred.

It is essential to plan and monitor this process. The details are too complex for inclusion in the high-level Capacity Report. It gives an overall view of the changeover process and shows how well the process is approaching the planned completion date. Any unreasonable deviations will result in further investigations at the local level.

10.5.5   The Capacity Forum

The Capacity Report would lose much of its worth without the Capacity Forum. There must be an interested readership ready to act on the results. Fortunately, the report is widely read by senior managers and there is no difficulty in promoting in-depth discussion. The report 'feeds' the discussion, but it is not the only feed. The forum existed first.

'Forum' is a good word, because it brings together managers with a wide range of different responsibilities. They discuss the capacity and performance issues (but not just those in the Capacity Report). Readers receive the report by e-mail, or obtain it from a protected Web site, so they are prepared for any discussion that ensues. The report is presented verbally for review and discussion item by item to ensure that nothing is overlooked.

10.5.6   Customer Satisfaction

Customer satisfaction is always in the minds of the Capacity Forum. The aim is to encourage more business, attract more customers and, of course, keep customers.

Adverse publicity is one of the problems that can arise from any kind of shortcoming. Failures may be sensationalised by the press, rather than explained logically. However, there may be opportunities to discuss issues with major customers, in moderately technical terms, and describe what is being done about them.

The standard of service of some major customers is reported, and other customers are identified if their standard of service has been threatened or compromised.

10.6   Summary

The Capacity Forum is at the centre of the whole capacity management process, while the Capacity Report is at the centre of the data collection and analysis. The Capacity Forum is the co-ordinating team that keeps the issues and methods under constant review. It recognises that focusing the reporting into one work area is efficient, economical and effective. It takes the statistical reporting task away from separate business units, which contributes to local morale. It puts the task under the control of a dedicated individual who is free to develop the skills required to produce a very high quality report.

The motivation for this emphasis on capacity and performance is 'customer satisfaction', but capacity and performance engineering is also about minimising costs. The aim, therefore, is to prime the business with the network information required to achieve customer satisfaction at minimum cost. Covering both systems and services, each new issue of the Capacity Report builds on the picture established in previous months, and projects this picture into the future. Managers use the report, primarily, for managing engineering intervention necessitated by forthcoming special events, system changes, new services and market trends. In addition, account managers of major customers appreciate the information delivered through the Capacity Report, because it puts them in a position of real knowledge to discuss issues that their customers may not fully, or correctly, understand. It also enables them to discuss new services, and how to create better business solutions from the current and future product range.

The TeleMarketing Services Capacity Report has been in production every month for several years. It was a very successful innovation that has grown and evolved in a way that could never have been envisaged at the outset. A web of links now reaches out to the business, giving the kind of coherence and structure that encourages objective discussion among otherwise disjoint communities. With its comprehensive and flexible coverage, straightforward and consistent format, simple graphs and tables, together with its emphasis on end-to-end performance and focus on customer satisfaction, the TeleMarketing Services Capacity Report stands as a prime example of how to monitor a complex, evolving system, and plan its progress towards providing the yet more sophisticated services of the future.

11

PERFORMANCE HEALTH CHECK

L G Kirby

11.1   Introduction

A full performance analysis of a complex computer system can be difficult to complete in a short time-scale if the system consists of many interworking sub-systems, often widely separated geographically, and if time and budget pressures prevent the normal in-depth assessment of performance. There is, however, a way to identify performance problems and provide indications of where performance engineering effort should be directed to greatest effect — the performance health check.

This chapter covers the full richness of the performance health check; normally a subset of the techniques would be employed, tailored to the particular requirements of the situation — time-scales often dictate how deeply the analysis should go. Note that the approach here concentrates on systems; networks would be covered in an analogous way.

11.1.1   What a Performance Health Check Is

A performance health check is a high-level, broad-brush survey of the performance characteristics both of the system and of the practices of the delivering project. The primary purpose of the health check is to identify key performance risks for a project. History has shown that major performance problems, or even complete project failure, could have been avoided if performance risks had been identified at an early stage and managed effectively. The health check approach can usefully be applied in a variety of situations:

- the project may still be in the early stages of design and development, it may be in the throes of deployment, or the system may have been deployed and operational for some time — a health check can still help to avert performance problems;

- the system may have suspected performance problems that need to be confirmed and summarised, or there may be no current issues and a demonstration of due diligence is all that is needed;

- the system may be a pair of simple subsystems which interact, or a complex configuration with, perhaps, thirty or more subsystems processing many different types of work — the health check approach will allow a single view of a complex system, and provide a strategy for mitigating performance risks.

An analogy is a medical examination for a person by a doctor — it is conducted at a fairly high level, with the intention of identifying symptoms of deeper problems, perhaps to provide confidence that an increased level of physical exercise is acceptable.

The entire end-to-end system is surveyed, as are the individual subsystems of which it is composed, and these are often computer systems in their own right. This survey analyses the performance characteristics and capabilities, and compares the results with what is required of the system and its subsystems. All of the business processes, and their related primary workstrings, which may be applied to the system are considered. As well as the performance capability of the delivered system, the health check process analyses the way that the system is designed, developed, and delivered; it also considers the way that the performance of the operational system is managed.

The result of the analysis is a performance health check report, which provides a summary of the health of the system, and an indication of further performance engineering work which might be advised in order to mitigate or eliminate the identified performance risks. It also describes the ways in which performance as a concept is considered in the project, and provides advice on strategies which could be adopted to the system's advantage. In short, it reports on the performance capability and the performance culture associated with the system; in this way a single report encapsulates the performance complexities of an interacting set of subsystems, as well as providing a performance engineering strategy for the system.

It is important to note that the health check is conducted at a high level — its scope is broad. All of the subsystems used in the primary workstrings are considered, requiring experience and understanding of a wide range of system types. A degree of independence in the performance analyst conducting the health check is also invaluable when problems arise and component teams are seeking the cause.

The health check should not be viewed as the final statement regarding system performance. For one thing, any changes to the system or the way it is used are likely to affect performance, and thereby require a new health check. And if the health check highlights performance problems, these might be masking deeper ones which are equally severe — in this sense, fixing performance faults can be like peeling the layers of an onion (with just as many tears).

11.1.2   What a Performance Health Check Is Not

Since the performance health check has to be high-level, certain activities normally expected in a performance analysis are not included. Examples are performance modelling, and system or subsystem performance tests. If, however, any of this has already been done by any of the teams working on the project, then obviously the results are used in drawing conclusions in the health check report — and the performance risks are affected accordingly.

The health check report will include pointers to performance problems, but not necessarily provide any solutions (although suggestions may be made); the emphasis is on a speedy analysis. To continue the medical analogy, the doctor may suggest areas requiring further analysis, or refer the patient to a specialist, but invasive diagnostic surgery is definitely not part of the health check.

A health check generally requires fewer technical activities than may normally be expected for a performance analysis; for example, much of the work is reading documentation, interviewing people, and taking informal measurements — as opposed to long-term monitoring, formal testing, or producing and using simulation models.

11.2   Health Check Activities

The approach to conducting the performance health check consists of understanding the following:

- workstrings, their volumes, and how they are processed by the system;
- elapsed-time requirements;
- instrumentation and monitoring capability;
- testing or other assessments (for example modelling, monitoring the live operational system) planned or already done;
- overall performance capability — how well it meets the above criteria.

This is done for the subsystems as well as for the entire end-to-end system. Any lack of information in the above categories leads to increased risk of poor performance in the deployed system.

These points are discussed in detail in the following sections.

11.2.1   Workstring Characterisation

Understanding the types of work the system has to handle is a fundamental part of any computer system performance analysis, and for completeness we include a description here.

This section describes how workstrings are identified, their associated workloads, and the elapsed time targets for the system to handle them.

The usual approach is to break down the sources of work, and describe the workstrings each source will submit to the system — typically these may be the following.

- Users

  There may be different types of user, and each submits a set of workstrings on the system. For example, there may be operators using their PCs to manage their interactions with customers, whereas other users may access the system once a week to log their time.

- Batch

  There may be different batch jobs, each being a separate workstring, for example, at midnight the system may 'wake up' and execute:

  — a billing run to take all customer data to produce bills for the previous period's service,

  — a database archive of jobs, or database reorganisation,

  — a download of data from another system (or a subsystem of the system under study) to refresh the database,

  — an update to the contents of a Web portal.

- System-to-system

  There may be transactions between subsystems, or between the system being assessed and another system. For example, an external system may request information from the system being assessed.

Table 11.1 shows a simple example of a high-level description of the workstrings that were considered in a health check for a call centre. In this case the hardware in scope for the health check processed only contacts from the Internet by a member of the public (denoted in the table as 'customer'), and workstrings not having an impact on those hardware components were deemed to be out of scope (shown with tint in the table).

The performance analyst must understand at least the high-level system design, but perhaps not so much of the internal detailed design of the subsystems.

Once the different categories of workstring are understood, the workloads of each are established; when this is specified in terms of an amount of work in a given time period it is often referred to as 'throughput'.

For analysis of future system behaviour, the forecast workloads over the next few months, or even years, should be established. The performance analyst will then be able to begin to assess the impact of the workstrings on the performance of the system.

Table 11.1 Sample workstring summary.

Process	Description
Inbound call	Customer calls with query, or a fault, or wishing to change account details
Outbound call	Agent calls customer about BT promotion or service
Call me — immediate	Customer clicks on button in borwser for agent to call immediately
Call me — scheduled	Customer clicks on button in browser for agent to call at some point in the future
Co-browse	Agent initiates a browser on customer PC, copied on agent's PC — initiated during text chat
Text chat	Customer and agent communicate via instant messaging
E-mail	Customer sends targeted e-mail to BT
Future opportunity	Agent initiates a call based on customer's previous interest

The emphasis in a performance health check is on timely production of the report; therefore it is important that the analyst takes care to consider only the primary workstrings, and ensures that the number of different types of workstring does not multiply beyond necessity. If two workstrings are similar in many ways, the analyst may decide to combine them into a single representative workstring; similarly if any workstring is insignificant from a performance point of view, then it is sometimes ignored completely.

For example, in a given operational support system, the most common interactions might be orders and 'cease service' requests, plus fault reports; however, there will also be order cancellations. If the orders, requests, and fault reports occur at a rate of more than 100 an hour every hour, yet the cancellations occur once an hour and involve only one or two of the front-end systems, then the analyst may decide that order cancellations may be ignored for now. This assumption should be checked with the customer.

Once the analyst has an understanding of the work which the system has to handle, the next step is to establish how quickly the different workstrings should be processed. Note that this is not the same as throughput: if the throughput is stated to be, say, '60 transactions a minute', then this means that, in a typical minute, 60 transactions will arrive at the system, and 60 transactions will be completed — but not the same 60 transactions.

The steady value for throughput would indicate that the system is stable, but would not mean that the elapsed time for an average transaction is one second; in fact, without further information, it says nothing about the elapsed time because the number of transactions waiting in the system is not known. Therefore, a stated throughput requirement for a workstring does not say anything about its required elapsed time; the elapsed-time requirements, therefore, also need to be established. Again, the performance engineer may decide to combine or ignore certain workstrings for expediency.

A full specification and thorough understanding of performance requirements is vital to any further performance analysis — and a full performance assessment is not possible without them. Elapsed-time targets should have been specified in the non-functional requirements documentation for the system.

The best way to specify an elapsed time target is at the business level. For example, how long it should take to handle a customer on the telephone at a call centre — while the individual components of the call (such as searching for customer details, updating the customer's account) are important, the overriding target should be the time that the agent is occupied.

Once this high-level elapsed time target has been specified, targets for its components can be established. For example, the elapsed-time targets for user interactions are normally expressed in terms of an on-line transaction response time, such as:

'When the user inputs the customer number and requests the customer address, the time taken between clicking on OK and the address being available should be less than 2 seconds for 95% of the time.'

The project may have decided to combine workstrings with respect to response time targets, as in, for example:

'For any user transaction where the user clicks on OK the required data should be on-screen within 2 seconds for 95% of the time.'

However, using a single requirement to cover a wide variety of situations in this way should be regarded with caution; it is open to misinterpretation, and is difficult to follow in practice, because it does not take into account the differing characteristics of the various workstrings.

For the health check report, the workstring information is gathered and summarised in sections describing:

- the purpose of the system (briefly);
- the primary workstrings applied to the system, their associated workloads, and how they move through the system;
- elapsed-time requirements for each workstring.

11.2.2   Individual Subsystem Capabilities

Now that the performance analyst understands the work the system has to do, how it is handled, and how quickly it should be completed, the next step of a performance health check is to understand what impact this work will have on the subsystems.

For each subsystem, the health check needs to establish the following information.

- Workload

  It must be determined how the volumes associated with each workstring which have been specified for the system as a whole (as determined in section 11.2.1) are broken down to apply to the subsystem.

- Elapsed-time targets

  It must be ascertained how quickly the subsystem should process the different workstrings — in other words, the performance requirements. These may be derived from an apportionment of the system-level elapsed-time targets.

- Instrumentation

  It must be established whether the subsystem is instrumented and what happens to the information, and how it is used. 'Instrumentation' means specific pieces of software, or sometimes 'hooks' in the code, which enable the system and the application to measure their own performance — and if the performance is not measured it cannot be managed. This instrumentation is commonly at either the application level or the system resource level — instrumentation at the application level usually refers to some sort of time-stamped log of application-level events, whereas instrumentation at the system resource level means the normal hardware-related monitoring such as CPU, memory, disk, and so on.

- Assessment

  The level of performance analysis already undertaken or planned must be ascertained. This includes performance modelling, testing, and monitoring of the live operation.

- Capability

  A simple statement must be obtained of how well the subsystem meets its expectations (basically a set of statements interpreting and summarising the previous few items) and how well it will achieve its targets in the light of the workload forecasts.

- Rating

  A red, amber, or green (RAG) score for the subsystem (one for each workstring) is noted here and repeated in the management summary of the report to bring all the RAG scores into one place; an example of a RAG summary table is given in Table 11.4 in section 11.2.4. Note that sometimes a 'blue' score is appropriate, when the system performs so well that it has probably been overprovisioned.

The report should contain a picture showing all the components in the system, to make it clear which subsystems are inside or outside the scope for the health check. It should also indicate those systems outside the scope but which play a part in normal operation. Figure 11.1 shows an example of a fictitious system architecture diagram.

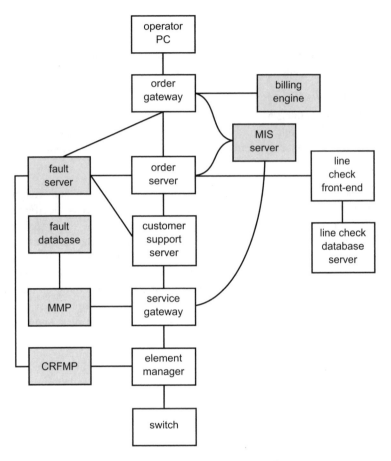

Fig 11.1   Sample system architecture diagram.

For each subsystem, the report should also include:

- a brief description of the subsystem (configuration, uses, processes), and the sources of information (including primary contact names) (see Table 11.2 for an example of a component description);
- an outline summary of the subsystem capability, as in the bullets above — this could be captured in a table if required, as shown by the example in Table 11.3;
- a list of the risks and issues found as a result of the health check.

11.2.3   System View

Having established the capability of the individual subsystems, attention now turns to the system as a whole and its approach to performance.

Table 11.2 Sample component description.

The customer database server runs on a Sun E45000 with 8 CPUs and 8 GB of memory, and two RAID arrays of 3 disks each (each disk 18.2 GB) plus a hot spare. This component processes requests for either new accounts, or changes to existing accounts, or account removals. Information about its performance was obtained from Anne Other and her team of developers.

Table 11.3 Sample component performance assessment summary.

Workload targets	Detailed volumetrics have been supplied to the developers for this component for the number of requests to receive in a typical day (10 000 requests), but not during a peak hour. The developers report that typically their component has, on other projects, experienced a peak hour which is 30% of the daily total. Based on this, they estimate that the peak hour number of requests will be about 3000; this component has been designed to meet this workload. Note that the requests are of two types: orders and modifications (deletions are so rare as to be negligible); about 80% of the requests are orders. There are no workload forecasts available for the next financial year, apart from a general consensus that they could increaxse dramatically.
Requirements	The designers were instructed that each request should be completed within 2 seconds average, with a maximum of 5 seconds; longer than this would be unacceptable.
Instrumentation	This component keeps a time-stamped flat file log of reqest entry and request completion time, so that response times can be determined. The results are published every Friday, after extraction and processing, in a simple report available on the corporate intranet.
Previous assessment	Performance testing was conducted on a previous release of the software, indicating that performance targets would easily be met; however, the current release has only been tested for orders. Note also that the previous testing indicated that modifications took slightly longer than orders.
Capability	The level of requests submitted to this component in the first few months of its life as part of this system will be negligible. After that, the volumes will begin to increase until they reach the speficied 10 000 a day. Results from testing the previous version of the software indicate that there will be acceptable performance in the current release for orders, but only a level of confidence for modifications. Since there are no formal workload forecasts for the next financial year, there is no confidence in the component for that period.
Rating for each workstring	Orders this FY:GREEN Order Modifications this FY:AMBER Orders next FY:RED Order Modifications next FY:RED

11.2.3.1 Instrumentation

This topic has already been considered for the subsystems, but it is also very important for instrumentation to be available for the whole system — as always, if performance is not measured then it cannot be managed.

The information is, again, usually at two levels — application and system resource. In practice, this means that the information available from subsystem instrumentation is gathered together, processed, and then made available to people who provide operational and application level support, as well as to performance engineers. Normally this is via an on-line presentation, or in a paper-based report; either way, a performance report on the overall system resource and application performance should be available regularly (every week or month, depending on the requirements) — see Chapter 9 for further discussion.

The health check report therefore describes:

- the end-to-end system monitoring capability — whether there is some measure of the overall end-to-end system performance (application and resource);

- what happens to the information, how poor performance is indicated, and how the operations people react to it.

11.2.3.2   End-to-End System Performance Assessment

This brings together any assessments of the entire system, including performance testing and modelling, which may have been done (or is planned) by a specialist team. A detailed description of an approach to performance modelling in this context is provided in Chapter 10.

Another activity included under 'assessment' is monitoring the live operational system — if the system is instrumented in any way, then measurements taken during its operation can provide further insight into the system capability. Sometimes this is done by gathering together all the subsystem monitoring to make it available from a central place so that an overall view is obtained.

The health check report therefore describes:

- performance testing and modelling, including who, what, and when — if testing or modelling has already been done, then the results can be used as part of the overall system assessment in the health check report;

- performance monitoring results from the operational system (obviously this will only be available if the system is live) — this data is used as part of the overall system assessment.

11.2.3.3   Risks and Issues

Next, the report should describe the current status of any previously known performance risks and issues. Normally, as a system is being designed and developed, the project begins to raise risks and issues, including those related to performance. The health check report should describe the project's approach to performance, including a description of the existing performance-related risks and issues and how they are being progressed. It may be useful to consider those which have been closed in order to decide whether the reasons for their closure are still valid.

11.2.3.4 Performance Management

Finally, the overall approach to performance management is assessed. This concentrates on the performance culture apparent in the project and in the people managing the system, and not the performance capability of the system. Some aspects of performance management have already been described above, but points to consider in the assessment include:

- whether the project has an understanding of the amount of work the system is expected to handle, whether the volumes have been forecast, and whether there are any actions planned to take account of the changes;

- whether the system design includes the ability for the application to report on its own performance (i.e. system-level resource monitoring and application-level monitoring), the quality of the information and how it is used for the benefit of system performance;

- whether there has been any activity to anticipate performance problems in any way, including performance modelling, performance testing, and performance risk assessment.

In short, the report must establish how much of a performance management strategy is in place for the system.

11.2.4   Overall Picture

Finally, the performance health check report summarises the findings from the health check, highlighting the areas where further work is recommended. A one-sentence summary then leaves the reader in no doubt about the results of the health check.

The report also contains a management summary of the information in the report, usually in the first section, which includes:

- a summary of the RAG scores for each subsystem, summarised in a table — Table 11.4 shows an example of a RAG summary table, which indicates that (for example) the Unix1 component can handle all of the workstrings apart from 'MIS reports', for which there are possible risks for the current workload, and definite problems for the increased workload expected next year;

- a summary of the performance risks and issues;

- the one-sentence summary of the performance health of the project.

Table 11.4 Sample RAG summary.

	Current					Next year				
	Orders	Nightly Data Feed	MIS Report	Bill Productions	Fault Handling	Orders	Nightly Data Feed	MIS Report	Bill Productions	Fault Handling
PC				N/A					N/A	
NT1				N/A					N/A	
NT2				N/A					N/A	
UNIX 1			▓					█		
UNIX 2										
Network	▓						▓			

Key: | OK | | Maybe | | NO |

11.3   Summary

A performance health check is a relatively quick way to identify the key performance risks in a project, and thence to provide an indication to management as to whether there is any need for them to take action (from a performance point of view). Sometimes it acts as a demonstration of 'due diligence' and 'best practice'.

The health check is important because it provides an indicator of where performance problems are, or might be in the future, and therefore provides a focus for where performance engineering effort should be directed to greatest effect. It can also make suggestions for strategies for the project to adopt and which would improve the performance culture. If the performance risks are not identified early and managed effectively, the project is likely to suffer from performance problems and may even end in complete failure.

There is no reason why the performance health check should only happen once to a system. In the same way as people can have annual medical examinations, the system can be subjected to a health check at regular intervals, or at other significant times in its life (for example, when the software is upgraded, or when performance testing results are available). In this way, management can feel confident that the health of the system is being monitored, assessed, and judged.

12

PERFORMANCE AND THE BROADBAND WINDOW

I D Pearson

12.1   Introduction

As early as 1985, the concept of broadband was already well established throughout the telecommunications industry, and was often applied to any service that needed 140 Mbit/s of bandwidth or more, though it soon came to mean a few tens of Mbit/s upwards. Anything less than 2 Mbit/s was definitely considered to be narrowband and 2-30 Mbit/s was called mid-band. That was mainly because video then needed 140 Mbit/s to get good quality, but video delivery was seen as very achievable in spite of its huge bandwidth appetite. Several systems were designed to carry tens of gigabits per second over local networks so that each customer could have over 140 Mbit/s. In the late 1980s, some high-density wavelength division multiplexing (WDM) systems were designed with the potential to deliver 2 Gbit/s to each home.

Later, some simple calculations, based on retinal cell performance characteristics, showed that 2 Gbit/s was the highest data rate that the human body's senses could handle. The ultimate broadband network would provide several channels of this capacity to each home to offer full sensory capacity to each home member and a few electronic devices. Even with a total access rate of around 10 Gbit/s, many telecommunications engineers felt confident that they could engineer such rates into networks by the late 1990s or even earlier. Yet, fifteen years on, we still do not have the 10 Gbit/s to each home that we had hoped for, anywhere in the world. Broadband today has become a marketing term that describes a share of 500 kbit/s. The telecommunications industry appears to have underachieved its original target by a factor of 20 000. So what is left of the superhighway dream?

There will still be a significant market for redefined broadband if the performance is good, but some of the originally expected market areas are reduced or even gone forever. Other markets remain intact, but no longer need broadband, since technology has enabled the same service to be realised using narrowband. Most of the technologies that have reduced the need for broadband have been developed by the telecommunications industry itself, for good reason. Some services that once would have needed a fixed fibre network could soon be achieved over a mobile connection. This means network operators do not have as great a need for a broadband network as they once did, but can offer the same high-quality services to customers at greatly reduced costs, and with much greater flexibility. But it also means that such services are much more open to competition. This chapter will take a critical personal look at traditional and future potential broadband markets, consider competing technologies, and try to argue whether or not any significant network market exists in each case. While there may be no single 'killer application' that will justify a high-speed network on its own, there are many services that add together to give traffic demand, and some services that can soak up whatever bandwidth is available, if the price is right.

Technology is only one potential driver for broadband services — social, political and environmental factors might also drive demand. For example, a high green tax on petrol might increase use of remote working services. However, this chapter will concentrate only on technical drivers, with section 12.2 covering the traditional broadband services and section 12.3 describing some emerging services.

12.2   Traditional Broadband Services

Firstly, let us look at some of the services that were considered in the early 1990s to be the main justification for broadband roll-out. High-speed distribution of music and video was considered a guaranteed winner and the emerging demand for e-mail and Web access were also considered very probable major contributors. Videoconferencing was often hailed as a major broadband service, but there was always a strong element of uncertainty about its likely success. These services have all evolved, and there are competing delivery systems too. The following sections will address each of them, looking at what is still left that can realistically be captured.

12.2.1   Network Back-ups

Network access data rate has certainly increased over the years. We have gone from 2.4 kbit/s modems in the early 1980s through to 56 kbit/s modems today. ISDN, DSL (digital subscriber line) and cable modems can offer hundreds of kbit/s to some customers. However, when we compare this progress with the rapid escalation of computer performance, we see that access networks have lagged far behind. This has meant that some potential network services are unfeasible. The 1990 traffic model showed that while telephony traffic would grow at a few per cent per year, electronic mail would grow at many times that rate until it eventually dwarfed voice traffic. But e-mail traffic would only generate roughly 7% of the traffic produced by back-ups. Most back-up traffic stays within a few metres of the desk. However, it is clear that if a network operator could build a broadband network with sufficient capacity, it might be possible to offer back-up services across that network. People could avoid worrying about where their back-ups were kept, and whether they were secure, and would even be able to access their files on secure links from other computer terminals anywhere. It could become a very valuable service.

However, as Table 12.1 clearly shows, the time it would take to back up all the files on a high-end PC has increased dramatically. In 1986, it would have taken a VAX 750 just 1.8 sec to transmit all my files to another machine over the Ethernet. By 2001, even if I had a 32 Mbit/s link, it would take almost 7 hours to back up all my files. The home network is even worse, going from a barely tolerable 2 hours in 1986 to 72 days now, even if I was to get a sustained 128 kbit/s on an ADSL line 24 hours a day. If these trends continue as expected, by 2010, I would be able to do only one complete back-up before it is time to change my computer again. So full network back-up could hardly be described as an attractive service. The full potential revenue from back-up services could never be achieved on any planned broadband network.

Table 12.1 The growing gap between computer and network performance.

	1981	1986	1991	1996	2001	2010
RAM	32 kb	500 kb	8 Mb	64 Mb	512 Mb	2 Gb
Computer speed	0.1 MIPS	0.5 MIPS	MIPS	200 MIPS	3 GIPS	300 GIPS
Agregate file volume	356 kb	2 Mb	80 Mb	2 Gb	100 Gb	10 Tb
Terminal access rate	—	10 Mbit/s	10 Mbit/s	10 Mbit/s	32 Mbit/s	100 Mbit/s
Time to back-up (LAN)	—	1.6 sec	64 sec	0.44 hr	6.9 hr	9.3 days
Telco line rate	—	2.4 kbit/s	9.6 kbit/s	28.8 kbit/s	128 kbit/s	2 Mbit/s
Time to back-up (local access)	—	1.9 hr	18 hr	6.4 days	72 days	66 weeks

Yet a back-up service is still feasible in spite of all this. It might not be easy to back up a whole hard disk, but customers will be able to make network back-ups of important files. The success of the service depends mainly on the ease of use and the speed of transfer. Each individual document I produce is only a few megabytes, which would not take long to upload even on ADSL. The key factor is convenience and ease of use. If we have a back-up of our applications and system files on a disk at home, we only need to worry about our documents, and these are produced one by one over a long period of time. If we run an application that automatically backs up each item we produce, immediately after it is produced, a high-speed network would easily be able to cope with this. Network back-up is only a problem if we try to back up large numbers of files at the same time.

It is perfectly possible to write software that interfaces seamlessly with a network to provide a secure back-up for our precious files. A very occasional back-up of the disk as a whole would still be sensible on a digital versatile disk (DVD), fluorescent multilayer disk (FMD), or whatever, but most networks could easily cope with an incremental everyday top-up. Similarly, restoring individual files could be quite rapid, and an FMD could be mailed if a large quantity of data needed to be restored. At least the data would be safe if our house were to burn down.

Network back-ups are therefore a significant potential contributor to broadband if the package is right. They can soak up almost any data rate that the network could provide, but the actual contribution to network revenue will depend on cost, transparency and performance.

12.2.2 Main Back-ups

Although networks will be useful for partial and incremental back-ups, they cannot compete when it comes to full back-ups, simply because of the huge volume of data involved. There are much faster ways of backing up our files using DVD writers and their future derivatives. A DVD can easily be sent by post or carried in a pocket too. It already looks likely that DVDs may soon be replaced by FMDs, and these could hold 500 Gbytes. With this sort of capacity on cheap hard media, it is unlikely that networks will ever be able to compete on cost and speed with local disk-based back-ups. A high-speed connection today could barely do the job, and, as the gap between network and computer performance is increasing, it will be even less feasible in future.

12.2.3 E-mail

E-mail generates far fewer total bits per second per user than back-up traffic, but those bits usually travel further. For most companies, the amount of internal traffic far exceeds that which travels on to the public networks, suggesting that there will always be a relatively strong market for broadband local area networks (LANs) and LAN interconnect services. However, some e-mail has to use public networks, and this accounts for an increasing proportion of total network traffic. E-mail has never been seen by users as a real-time service, and LAN performance is perfectly adequate. DSL will also offer adequate capacity and speed for e-mail. Only if video content really takes off will such networks be challenged. However, public switched telephone network (PSTN) modem speed (56 kbit/s) is only adequate for e-mail without large attachments. E-mail has certainly shown its potential in business and has equally valuable potential social uses too, but carriers without high-speed networks will not be able to capture this potentially lucrative service. As computers become more powerful, and particularly as we see video and digital photos take off and increase in file size, domestic networks will struggle to keep up.

Social use of e-mail remains restricted because many people still do not have Internet access at home. Since many people are as yet unaccustomed to the delays that the rest of us have learned to cope with, their expectations may still be set by the instant response of other everyday equipment. Achieving mass-market penetration will mean reducing network delays, and also making the services easy to set up and use. The industry cannot afford to be complacent about acceptable performance levels, even for services such as e-mail.

For many people, e-mail has also evolved into instant messaging. Many social groups stay in constant touch with each other via instant messaging and text messaging on telephones. We expect this to increase in the future, and also that instant voice and video messaging will evolve — which will obviously make much higher demands on capacity.

12.2.4 Web Access

The 1980s' idea of a global superhighway, with lightning fast access from homes to data and services from all over the world, was hijacked in the early 1990s by the Internet. When the World Wide Web was invented, the two concepts quickly merged and the Internet was seen as a poorly designed not-so-super-highway that could be patched up over time to finally realise the superhighway dream. Sadly, many services that could have flourished on a superhighway could not work on a low-performance Web. Although 'dot-coms' failed for a variety of reasons (many were based on highly flawed business models and a complete lack of understanding of the nature of the Internet), I believe many were perfectly viable businesses that could have succeeded if access was fast enough and easy enough. They were simply launched on to a network that was not ready. When it takes ages to connect, dial up, wait for authentication, and then even longer to connect to sites and download data, it is no great surprise that people mostly did not bother. Human factors tests in the mid 1980s suggested that people ideally need a response within two seconds to a complex enquiry on a computer terminal and almost instant response to a simple button press.

Obviously this depends a lot on context, but those recommendations were for response on a VT100 terminal, where people are accessing material held centrally — not so different conceptually from what the Internet offers. That kind of response is rare on the Internet today, even when the connection is fast. Our current Internet performance expectations are set by our experience of using an impaired system, rather than on what is desirable.

Computer systems are often overloaded and links frequently swamped with traffic. Many dot-com ideas will rise again and flourish once most people have fast access. Moving to 'always-on' connection has already removed log-on time for many people.

Improving the performance of network switches, using storage to spread and reduce traffic, and improving server performance, will all work together to achieve the two-second goal. Only then will we discover whether there really is a viable dot-com market.

12.2.5   Web via Local Storage

Internet access is a fast-growing market in spite of many of the dot-coms collapsing. But most Internet traffic is to just a few sites, and the proportion of new sites that are of any real interest to a significant number of people is small. Many sites that people access have data that changes frequently, particularly on front pages, but in most cases, just a few articles change, while the vast bulk of data is in slowly changing archives. Relatively little of the data downloaded is truly dynamic. It is possible that Internet magazines could include the key content of most of these new sites on the DVDs or FMDs stuck on their covers. A home store could extract the data from the FMD that is likely to be of interest to the household members, according to their filter profiles (Fig 12.1). Since it only keeps what is likely to be of interest to us, the home store could thus contain most of the relevant information from the Internet. This is possible because of the rapid growth of storage capacity on hard disks. In just a few years, multi-terabyte capacities will be the norm. This is such a large amount of data in human terms that it could reduce our need to directly access the Internet dramatically. Retrieving information from a local hard drive will seem utopian in speed terms compared to doing so across a slow local access. The machine will only have to cope with one or two simultaneous users and network latency will be tiny since the user will be in the same house. Once this market grows significantly, it will be difficult to lure people back on to the access networks. Since the Internet is global and the speed of light finite, even with the best possible technology it will be relatively slow accessing data across the Internet compared to

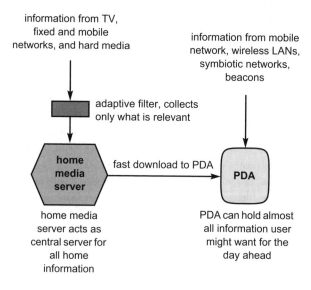

Fig 12.1   Storage-based networking.

this local access. We may always want to keep most of the relevant data from the Web on a local store, and only access the Internet for the remainder — a small fraction of our total use. Performance of network-based access will always be poor when compared to using a home store.

However, many people will prefer to put up with delays than to take the trouble to load FMDs into their server, or to teach the device about their preferences. I have no doubt that the two alternatives will live comfortably side by side. Although we could have much faster performance by using carefully managed local stores, we will trade some of this performance for convenience and the reduced need to plan. The network will capture a lot of traffic in this way.

The main application for the network in this respect may be using the free capacity to broadcast popular sites, thus reducing the need to distribute data on disks. Disks may have enormous capacity, but the number of sites people visit is so small that even a modestly high-speed network can deliver these during quiet periods. The network access will check the home store first to see if the site is cached and then use the Internet if it is not. Between accesses, it will absorb relevant data off the broadcast network.

In this way, we can use an efficient compromise between hard and networked distribution, giving the bulk distribution of hard media, and the responsiveness of local hard storage, with the timeliness of networks. We could have a pleasant mix of performance and convenience.

12.2.6   Music Distribution

In Fortune magazine [1], a chart illustrated an expected revenue from music of $20 per month per customer over broadband. However, network-based distributors will have to do well indeed to achieve this in competition with hard media distribution. A holographic disk or an FMD with a half terabyte of capacity (750 CD equivalents) could easily hold the entire output of any pop group, and would allow music producers to deliver a large part of their catalogue on a single disk, with users buying access keys across the Internet. Instead of downloading gigabytes, customers would just download a few bytes of keys to access the gigabytes off a disk.

This distribution model is already being evaluated by the industry, and the obvious security concerns can probably be solved. In spite of such competition, network downloading will undoubtedly take some of the market. People will sometimes want quick access to a new track, or want to browse through material they have not heard before, or follow recommendations on Web sites. Disks may hold some of the material, but not all. And we may not want the hassle of loading the right disk into our home media servers, when we can just click a mouse or the TV remote. High-capacity disks will win some of the market, but not all. There will always be a healthy network market for music. Whether it is $20 per month for each customer remains to be seen.

12.2.7   TV and Video Distribution

With such high disk capacity, even video distribution could be affected. The suppos-edly natural platform for digital TV distribution would be a truly broadband network to the home. Now that satellite distribution has gained such a dominance, land-based networks will find it hard to capture a share of this, even over fibre to the home. Satellite distribution will have to compete against both networks and FMD video distribution. This competition is likely to be accelerated by the move towards the use of digital video recorders (DVRs) or home media servers. As people get used to the facility, and as these machines get better at profiling their owners and anticipat-ing their desires, we can expect some psychological convergence between video and TV use. Having TV programmes appear on FMDs will then not seem so strange. Some customers might buy a whole year's subscription to a content provider's output direct instead of paying premium channel subscriptions to satellite distributors. The material could appear in the post instead of over a satellite link. We could each have a copy of every film in the video store on disk, enabled by a simple key download.

Although FMD has some big advantages, it will not have it all its own way. Satellite distribution will not be easy competition. Much of the TV that people watch is serialised, or relates to current affairs, so cannot easily be adapted to work with posted media.

There is a lot of scope for bandwidth saving on satellite distribution. Many channels on British satellite TV are currently used for 'Box Office', which is a staggered film broadcast. With good DVRs, which the satellite distributors are now pushing, a single broadcast of the film can be used, with customers loading it off the disk at their convenience. The same could apply to repeats, which currently take up a lot of channel space. Many other channels are part-time. Variable bit rate encoding can liberate some capacity on each channel too.

Amalgamating these spare capacity sources could allow extra material to be broadcast to DVRs at no extra cost compared to today. Since the DVR does not care whether video data arrives in real time, it can be delivered at whatever data rate is currently free, sometimes arriving as a trickle, sometimes in a flood. DVRs might receive most of their daily top-up during night-time quiet periods. If and when analogue signals are switched off and the bandwidth re-assigned to digital TV, this will become an even more attractive delivery means. It is debatable whether there will be enough content to fill the satellite channels without using repeats. If not, then some free capacity might appear that could be used to broadcast corporate videos, e-mails, software and other data.

It is hard to see how network distribution can compete with the combined might of these two approaches, unless it offers low cost and high bandwidths. Certainly, it will be some time before operators can profitably send 4 terabits across a network for 50p! It would take 740 hours at 1.5 Mbit/s, almost exactly a month. Even the postal service is better than that. If a daily FMD is to be beaten by a network, the network will have to operate faster than 46 Mbit/s, continuously. Fortunately, networks do not have to compete on those terms. BT's network has key advantages compared to a dumb broadcast distribution medium. With good marketing and swift roll-out of DSL technologies, BT may yet persuade some people to use their networks for their TV some of the time. Trials of video-on-demand over such networks have been successful, and, in spite of cost, some people seem willing to use it. BT has the advantage of having an individual channel to each customer, so can personalise offerings and offer much better interactivity than the other approaches can achieve. As television becomes more interactive, this advantage will increase. Even with conventional TV, the BT network could allow people to access channels from other parts of the world, rather than the walled garden provided by their satellite or cable operator.

12.2.8   Videoconferencing

Videoconferencing apparently rose dramatically in popularity after September 11th 2001. Many international conferences were cancelled as business people avoided non-essential travelling. However, people are now getting back to business as usual. We should expect that travel will once again take over from videoconferencing as the preferred way of meeting unless terrorism on aeroplanes becomes a regular threat. Videophones have also proved rather less popular than originally expected. So it might look like video may have been overestimated as a potential broadband market. However, I do not think there is sufficient evidence yet to make such a claim. Today, the number of people with terminals is low, the cost is high and the ease of use is lacking. All of these need to be put right before videoconferencing can be properly evaluated. Now of course, as the data rates needed for acceptable conferencing are very low, it is not technically a broadband service any more.

Future conferencing using avatars in virtual environments looks even less demanding for the network. This means that traffic per session will be lower, but it might mean that people would be far more likely to use it, since it will be available to them more of the time at acceptable cost.

Two technical problems in videoconferencing today are eye contact and encoding delay, and both of these could be addressed using avatar technology, or even a blend of video and avatars. Since avatars can also add a touch of fun to an otherwise dull communication link, progress in this field might yield rich rewards, especially since these technologies are also applicable to other fields such as man/ machine interfacing, games and leisure activities.

12.2.9   Computer Games

There is some potential for providing game playing across the network — it is already a significant market. People now download games and play against each other across the network. This is likely to continue, and broadband will obviously make such downloads and interaction faster. As games become more sophisticated and make more use of multimedia, broadband access may become essential to get acceptable download speeds. However, while games can certainly make good use of available bandwidth, the ability to access games at low cost via conventional media puts a limit on the potential price for such services. As games designers have also managed to achieve good interactive game play over narrowband links, broadband is very much an optional extra as far as play performance is concerned. Where broadband may prove useful in a games context is in permitting a continuous high visual link between the players. Beating opponents can be much more satisfying when you can see the expression on their faces. In fact, this might well be one of the ways in which video communications could become socially established. So even though game play may not actually need broadband itself, it may act to accelerate take-up of other broadband services.

12.3   Emerging Broadband Services

Although deployment delay may have shrunk some of the traditional broadband markets, there remain some traditional areas where network delivery will be preferred, if it is fast enough. The following sections look at emerging service areas that may contribute significant traffic. There are many new services that will add traffic to our networks, but I will address just a few with quite different characteristics — i-Com, cyberspace tunnels, personalised TV, real-time entertainment, community networking and smart environments. Although mostly they can get by with slower networks, they would all work better with higher rates and lower delays, and there will still be a market incentive, therefore, to improve performance.

12.3.1   i-Com

Videoconferencing may not be the killer application it was once hoped to be, and may probably never be a truly broadband service, but there is probably still a significant market for it if it can be made to deliver real improvements in quality of life.

A joint BT and MIT project, called i-Com, provides enhanced videoconferencing over IP. It is packet switched, and can use whatever spare capacity is available on a best-effort basis, and so the quality of the connection generally changes according to load, unless a guaranteed quality of service has been agreed. Another project, called Reflex, links electronic tags to network-based functionality. Combining these two technology areas with positioning systems could make videoconferencing more intuitive and more natural. However, performance must also be made much better, so that encoding and transmission delays are reduced, while picture quality must be improved by increasing bandwidth. Cost must be kept low too, or conferencing will remain a business tool. People working as part of virtual companies will find tools such as these will help them significantly to keep in contact with their colleagues. Similarly, people could stay in touch with friends and family when they move to another area. The low-bandwidth voice communications today contribute to people gradually losing contact with friends when they move, and they can become less close to other family too. This could be because voice simply does not convey the same emotional richness of face-to-face meetings, since body language is missing. If visual communication is easy to set up, people may stay in touch more frequently and with a higher emotional reward. But it is clear from market failure that a tiny picture that refreshes just a few times per second is not adequate. VHS quality is probably more than acceptable for most personal communications purposes in the near future, though larger screens will undoubtedly increase the requirement in due course. However, long compression delays will not be tolerable, as people expect their communications to be near real-time. But getting 'acceptable' picture quality with low compression delays will increase the data rate and cost. Offsetting this increase, expectations might be kept low by what people have to accept on lower capacity media such as mobile, and the increasing use of computer-based video communication using low-quality cameras and narrow channels. Here again, delaying broadband deployment will just make people more accustomed to accepting lower quality pictures, and make it harder to market higher quality later on. In order to soak up anything like a broadband rate, a single channel of videoconferencing would need to be high quality and large image size, preferably with lots of extra bells and whistles. However, it may be possible to justify several simultaneous channels to the customer and achieve a broadband demand in aggregate.

12.3.2   Cyberspace Tunnels and Virtual Sleepovers

My 8-year-old daughter sometimes has friends to stay for sleepovers. She would be delighted if I provided her with a large screen in her den, acting as a cyberspace tunnel to her friends' rooms. The children could chat and play games almost as well as if they were together. They could share a virtual sleepover every night. This type of call could easily last hours. A similar service might be useful in the kitchen, to extend mealtime gatherings into friends or relatives' homes. i-Com does something similar for business groups — my own group effectively has an occasional cyberspace tunnel through to an office in MIT. i-Com can use a few tricks to keep the performance needs down. It detects when there is any activity and then increases the frame rate. For real-time communication, the coding time could be reduced at the expense of higher data rate when the parties are actually interacting, relaxing to longer delays and lower rates when activity drops. I can easily imagine all-day links in telework centres keeping people in visual contact with their virtual company co-workers.

Future computing technology for videoconferencing will replace people by avatars for some of the time. The standard of realism in recent films such as Final Fantasy would be very acceptable, but it will be many years before such high-quality images can be rendered by home computers in real time. People could then change their appearance at will for whichever type of conference they are involved in. The application may well become much more popular but its performance requirements will drop dramatically, since the work will be done mostly at the terminals. This illustrates why communications companies need to focus more on adding value to communications rather than simply transporting bits.

12.3.3   Personalised TV

Personalised TV is a new entertainment category (Fig 12.2). On a broadband network, it will be possible to deliver individually tailored TV streams to each customer. They might watch the same programmes at different times, or they might have the characters or content of those programmes personalised in some way. Advertisements could be targeted at individual customers, and we may in the future be able to choose which presenter we want to present a programme. A personalised news programme might have our favourite presenter reading those news items in which we are most interested. However, to do this at the source would require very high computing power and even more to distribute all the channels. A much better way to do it in cost and performance terms might be to broadcast the text of the news to our home servers and have them create the video for us using avatars, voice

Fig 12.2   Interactive personalised TV.

synthesis, and so on. The same could apply to digital presenter substitution in chat shows and competition shows. It might be much better to do substitution at destination rather than at source. If so, then we might find that personalising TV allows some of the content to be stripped out at source, with compressed signalling replacing high-bit-rate video. Personalised TV could eventually be less demanding of a network than conventional broadcast. We should also consider the possibility, as we go further down the road, that interactive virtual environments will gradually capture time-spend from TV too, so that even less will need to travel across the network.

However, in spite of the future potential to reduce network loading, this trend would make good use of technologies such as avatars and virtual environments.

12.3.4   Real-Time Entertainment

Some entertainment is real-time, particularly sports, dial-in interactive TV and so on. It is sometimes unacceptable even to have a couple of seconds delay. If the neighbours receive the pictures of a goal earlier, they could be cheering before you even know there has been a goal! There should be a small market for technologies that ensure that all the networks deliver these services at the same time. However, such conditions only apply to a small proportion of all programmes. The vast majority of even interactive entertainment broadcasting is relatively insensitive to delays, provided they are below a few seconds. In fact, introducing small delays into networks can be a tool to help distribute interactive calls, but obviously fairness dictates that viewers on one network should not be permanently disadvantaged compared to viewers on alternative networks.

12.3.5   Social and Community Networking

Community networking offers a range of potential new services. Neighbourhood watch cameras can be networked so that people can see what is going on, and intruders are filmed and recorded in case of crime. The use of web-cams could allow people to attend meetings remotely, providing a low-performance conferencing service. Local TV channels may thus make good social use of spare capacity, but will not be a major contributor to commercial broadband markets.

12.3.6   Smart Environments

The future environment is likely to be peppered by large numbers of very small and very simple devices, all around us, even in packaging and in the paint on our walls [2]. Devices will talk to each other in very simple lightweight protocols, while a few of these devices will be able to interface to the larger world and act as relays. Processing will also use a just-enough methodology, which saves on energy and size.

With such ubiquitous sensors and processors, the network is very likely to be used for sensory collection, and realisation of a smart environment. Sensory collection will make our lives easier in many ways. Via networked video cameras, we will be able to see how busy it is in town, or what the weather is doing at our destination, before we decide when to go out. Audio sensors could monitor our conversations in order to anticipate our needs, proactively offering services according to the context.

The combined traffic from sensors and their supporting activators in our environment might be significant, but the technology is too embryonic to make any meaningful guesses about their performance requirements.

Linking these sensors, activators, cameras and monitors to processors, stores and filters will produce a very smart environment indeed (Fig 12.3). If we make the assumption that it will be mainly self-organising, and will make good use of both storage-based networking and peer-to-peer working, then it should be the case that traffic will be abundant, but mostly local. For efficient working, communication set-up times will have to be very low, which implies that distances will be small and that any additional network latency from protocol or switching delays will be very low indeed.

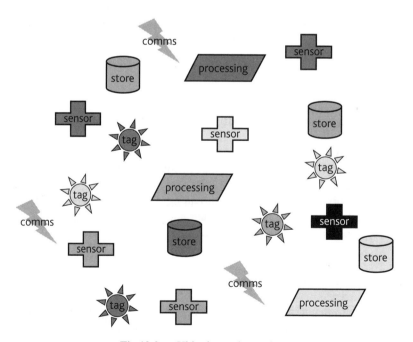

Fig 12.3   Ubiquitous electronics.

Based on little more than intuition, I suspect that message frequencies between devices will be very high, and performance will suffer greatly if total messaging time is more than a few microseconds. Probably the most suitable network would be self-organising symbiotic networks, using connectionless communication.

12.4   Summary

The dot-com collapses taught us, among other things, that networks are not yet ready for such companies to thrive. Performance must be sufficient to deliver fast access, fast response and fast download, or most customers simply will not bother. We do not have anything like the global superhighway that we dreamt of in the late 1980s. Many of the services that were initially dreamt up for the global superhighway have now been targeted by alternative distribution channels, and technology has reduced the necessary data rates for some of the others. Marketing has redefined broadband — so at least the term survives, even though the rates are what used to be called narrowband. A few high-rate services, such as computer networking, e-mail, file transfers, etc, may flourish on local area networks where bandwidth is abundant, but are not lucrative enough to justify the costs of a high-speed public network on their own. They will use processing power and intelligence in the terminals to make the best of whatever rates the network delivers. For long-distance communication of large files, customers may simply opt for hard media distribution, which may be cheaper and just as fast. Without the old killer applications, investment in broadband networks will be harder to justify. There will be rapidly growing competition from other information delivery platforms, making major investment much riskier. The sceptics who never thought there would be a broadband market, and therefore delayed its deployment, have partly achieved a self-fulfilling prophecy.

But the broadband dream is not dead, perhaps just a little more conservative. Although some of the original markets for broadband have shrunk, a number of markets remain for them, and new services have appeared that will make up some of the deficit. Broadband deployment may not be as lucrative as it might have been, but it will still be worthwhile.

The main highlight is computer communication. We will see increasing use of messaging, both instant and e-mail, and the richness of these communications will see message sizes rocket. This can only be acceptable if networks become much faster, and keep pace with growing computer performance. Of course, operators may not be able to justify investment on the back of one service type, but without good network performance, services will simply be held back from achieving their potential.

Most services in the future will not need high capacity, but will work much better if it is available — so their markets might improve dramatically with network performance. Since there are many services that will generate medium traffic levels, we should conclude that overall, there is still a strong need for a broadband network to carry the load.

Since risk/reward trade-offs will ensure that network performance will never be ideal, operators should continue to develop products that use processing and storage to offset the need for network transmission. In this way, they can reserve network capacity for services that need it, and also provide high quality most of the time for services that are somewhat less demanding and can survive on a best-effort basis. The few services that really need high speeds will simply have to wait for technology costs to come down, but by making imaginative use of other technologies, we in the telecommunications industry will accomplish at least some of our original superhighway dream.

References

1   Fortune magazine (4 Feb 2002).

2   Cosier, G. and Whittaker, S.: '*Reconnecting bits and atoms*', BT Technol J, **19**(4), pp 77-97 (October 2001).

INDEX